T0275486

Machine Tool Practices

Machine Tool Practices

Edited by
Raymond Foster

Larsen & Keller
www.larsen-keller.com

Machine Tool Practices
Edited by Raymond Foster
ISBN: 978-1-63549-168-5 (Hardback)

© 2017 Larsen & Keller

⊟ Larsen & Keller

Published by Larsen and Keller Education,
5 Penn Plaza,
19th Floor,
New York, NY 10001, USA

Cataloging-in-Publication Data

Machine tool practices / edited by Raymond Foster.
    p. cm.
Includes bibliographical references and index.
ISBN 978-1-63549-168-5
1. Machine-tools. 2. Machining. 3. Tools. 4. Metal-working machinery.
I. Foster, Raymond.
TJ1185 .M33 2017
621.902--dc23

The publisher's policy is to use permanent paper from mills that operate a sustainable forestry policy. Furthermore, the publisher ensures that the text paper and cover boards used have met acceptable environmental accreditation standards.

Printed and bound in the United States of America.

For more information regarding Larsen and Keller Education and its products, please visit the publisher's website www.larsen-keller.com

Table of Contents

VI   Contents

Preface

The aim of this text is to provide students with the most important concepts and methods of machine tool practices. It is designed specially for students to make them understand the basic and most complex techniques of this field. Machine tool refers to machines that are used for shaping and cutting heavy metals and rigid materials. It is used to grind, shear, cut and bore the metals. The topics covered in this book offer the readers new insights in the field of machine tool practices. It explores all the important aspects of this subject in the present day scenario. Coherent flow of topics, student-friendly language and extensive use of examples make this textbook an invaluable source of knowledge.

Given below is the chapter wise description of the book:

Chapter 1- Machine tools are tools that are used in shaping the metals used in machines. They are not used only for metals but for any rigid material that is different to cut or to bend. This chapter will provide an integrated understanding to machine tools.

Chapter 2- Broaching is the process that is uses broach to remove material. The two types of broaches are linear and rotary. Some of the types of machine tools discussed within this section are drills, power tools, hand tools, shapers and four-slides. The topics discussed in the section are of great importance to broaden the existing knowledge on machine tools.

Chapter 3- The processes and techniques used in machining are milling, grinding, abrasive machining, abrasive jet machining, productivity improving technologies, rapid prototyping and electrical discharge machining. Milling is the process that is used as a cutter in order to remove materials from any work piece. The aspects elucidated in this chapter are of vital importance, and provides a better understanding of machining.

Chapter 4- Loom is used in weaving cloth and tapestry; the purpose of any loom is to basically hold the warp threads in order to simplify the weaving of the threads. Lancashire loom, rapier loom and reed are some of the topics discussed in the following section. The chapter is an overview of the subject matter incorporating all the major aspects of weaving machinery.

Chapter 5- A machine-tool dynamometer is a dynamometer; it is used in measuring forces during the use of machine tools. The essential aspects of machine tools elucidated in the section are machining vibrations, machinist calculator, multimachine, ASME B5 etc. The section serves as a source to understand the major aspects related to machine tools.

Chapter 6- Numerical control is the automation of any machine tool, it can be operated precisely by commands and the storage medium is usually a computer. Some of the aspects of the numerical control are digital modeling and fabrications, cutter locations, Cartesian coordinate robots, CNC routers and CNC plunge millings. This section helps the readers in developing an in depth understanding of the subject matter.

Chapter 7- One of the most important technological developments of recent human history is the Industrial Revolution. It greatly developed the mechanization process, and one of the important inventions during the Industrial Revolution is the steam engine. Steam engines were not only used in the mining industry but also in various other industrial settings. Interchangeable parts, paper machine and coal mining are the topics discussed in the following chapter.

At the end, I would like to thank all those who dedicated their time and efforts for the successful completion of this book. I also wish to convey my gratitude towards my friends and family who supported me at every step.

Editor

Introduction to Machine Tool

Machine tools are tools that are used in shaping the metals used in machines. They are not used only for metals but for any rigid material that is different to cut or to bend. This chapter will provide an integrated understanding to machine tools.

A machine tool is a machine for shaping or machining metal or other rigid materials, usually by cutting, boring, grinding, shearing, or other forms of deformation. Machine tools employ some sort of tool that does the cutting or shaping. All machine tools have some means of constraining the workpiece and provide a guided movement of the parts of the machine. Thus the relative movement between the workpiece and the cutting tool (which is called the toolpath) is controlled or constrained by the machine to at least some extent, rather than being entirely "offhand" or "freehand".

A Metal lathe is an example of a machine tool

The precise definition of the term *machine tool* varies among users, as discussed below. While all machine tools are "machines that help people to make things", not all factory machines are machine tools.

Today machine tools are typically powered other than by human muscle (e.g., electrically, hydraulically, or via line shaft), used to make manufactured parts (components) in various ways that include cutting or certain other kinds of deformation.

With their inherent precision, machine tools enabled the economical production of interchangeable parts.

Nomenclature and Key Concepts, Interrelated

Many historians of technology consider that true machine tools were born when the toolpath first became guided by the machine itself in some way, at least to some extent, so that direct, freehand human guidance of the toolpath (with hands, feet, or mouth) was no longer the only guidance used in the cutting or forming process. In this view of the definition, the term, arising at a time when all tools up till then had been hand tools, simply provided a label for "tools that were machines instead of hand tools". Early lathes, those prior to the late medieval period, and modern woodworking lathes and potter's wheels may or may not fall under this definition, depending on how one views the headstock spindle itself; but the earliest historical records of a lathe with direct mechanical control *of the cutting tool's path* are of a screw-cutting lathe dating to about 1483. This lathe "produced screw threads out of wood and employed a true compound slide rest".

The mechanical toolpath guidance grew out of various root concepts:

- First is the spindle concept itself, which constraints workpiece or tool movement to rotation around a fixed axis. This ancient concept predates machine tools per se; the earliest lathes and potter's wheels incorporated it for the workpiece, but the movement of the tool itself on these machines was entirely freehand.

- The machine slide, which has many forms, such as dovetail ways, box ways, or cylindrical column ways. Machine slides constrain tool or workpiece movement linearly. If a stop is added, the *length* of the line can also be accurately controlled. (Machine slides are essentially a subset of linear bearings, although the language used to classify these various machine elements includes connotative boundaries; some users in some contexts would contradistinguish elements in ways that others might not.)

- Tracing, which involves following the contours of a model or template and transferring the resulting motion to the toolpath.

- Cam operation, which is related in principle to tracing but can be a step or two removed from the traced element's matching the reproduced element's final shape. For example, several cams, no one of which directly matches the desired output shape, can actuate a complex toolpath by creating component vectors that add up to a net toolpath.

Abstractly programmable toolpath guidance began with mechanical solutions, such as in musical box cams and Jacquard looms. The convergence of programmable mechanical control with machine tool toolpath control was delayed many decades, in part because the programmable control methods of musical boxes and looms lacked the rigidity for machine tool toolpaths. Later, electromechanical solutions (such as servos) and soon electronic solutions (including computers) were added, leading to numerical control and computer numerical control.

When considering the difference between freehand toolpaths and machine-constrained toolpaths, the concepts of accuracy and precision, efficiency, and productivity become important in understanding *why* the machine-constrained option adds value. After all, humans are generally quite talented in their freehand movements; the drawings, paintings, and sculptures of artists such as Michelangelo or Leonardo da Vinci, and of countless other talented people, show that human freehand toolpath has great potential. The value that machine tools added to these human talents is in the areas of rigidity (constraining the toolpath despite thousands of newtons (pounds) of force fighting against the constraint), accuracy and precision, efficiency, and productivity. With a machine tool, toolpaths that no human muscle could constrain can be constrained; and toolpaths that are technically possible with freehand methods, but would require tremendous time and skill to execute, can instead be executed quickly and easily, even by people with little freehand talent (because the machine takes care of it). The latter aspect of machine tools is often referred to by historians of technology as "building the skill into the tool", in contrast to the toolpath-constraining skill being in the *person* who wields the tool. As an example, it is *physically possible* to make interchangeable screws, bolts, and nuts entirely with freehand toolpaths. But it is *economically practical* to make them only with machine tools.

In the 1930s, the U.S. National Bureau of Economic Research (NBER) referenced the definition of a machine tool as "any machine operating by other than hand power which employs a tool to work on metal".

The narrowest colloquial sense of the term reserves it only for machines that perform metal cutting—in other words, the many kinds of [conventional] machining and grinding. These processes are a type of deformation that produces swarf. However, economists use a slightly broader sense that also includes metal deformation of other types that squeeze the metal into shape without cutting off swarf, such as rolling, stamping with dies, shearing, swaging, riveting, and others. Thus presses are usually included in the economic definition of machine tools. For example, this is the breadth of definition used by Max Holland in his history of Burgmaster and Houdaille, which is also a history of the machine tool industry in general from the 1940s through the 1980s; he was reflecting the sense of the term used by Houdaille itself and other firms in the industry. Many reports on machine tool export and import and similar economic topics use this broader definition.

The colloquial sense implying [conventional] metal cutting is also growing obsolete because of changing technology over the decades. The many more recently developed processes labeled "machining", such as electrical discharge machining, electrochemical machining, electron beam machining, photochemical machining, and ultrasonic machining, or even plasma cutting and water jet cutting, are often performed by machines that could most logically be called machine tools. In addition, some of the newly developed additive manufacturing processes, which are not about cutting away material but rather about adding it, are done by machines that are likely to end up labeled, in some cases, as machine tools. In fact, machine tool builders are already developing machines that include both subtractive and additive manufacturing in one work envelope, and retrofits of existing machines are underway.

The natural language use of the terms varies, with subtle connotative boundaries. Many speakers resist using the term "machine tool" to refer to woodworking machinery (joiners, table saws, routing stations, and so on), but it is difficult to maintain any true logical dividing line, and therefore many speakers are fine with a broad definition. It is common to hear machinists refer to their machine tools simply as "machines". Usually the mass noun "machinery" encompasses them, but sometimes it is used to imply only those machines that are being excluded from the definition of "machine tool". This is why the machines in a food-processing plant, such as conveyors, mixers, vessels, dividers, and so on, may be labeled "machinery", while the machines in the factory's tool and die department are instead called "machine tools" in contradistinction. As for the 1930s NBER definition quoted above, one could argue that its specificity to metal is obsolete, as it is quite common today for particular lathes, milling machines, and machining centers (definitely machine tools) to work exclusively on plastic cutting jobs throughout their whole working lifespan. Thus the NBER definition above could be expanded to say "which employs a tool to work on metal *or other materials of high hardness*". And its specificity to "operating by other than hand power" is also problematic, as machine tools can be powered by people if appropriately set up, such as with a treadle (for a lathe) or a hand lever (for a shaper). Hand-powered shapers are clearly "the 'same thing' as shapers with electric motors except smaller", and it is trivial to power a micro lathe with a hand-cranked belt pulley instead of an electric motor. Thus one can question whether power source is truly a key distinguishing concept; but for economics purposes, the NBER's definition made sense, because most of the commercial value of the existence of machine tools comes about via those that are powered by electricity, hydraulics, and so on. Such are the vagaries of natural language and controlled vocabulary, both of which have their places in the business world.

History

Forerunners of machine tools included bow drills and potter's wheels, which had existed in ancient Egypt prior to 2500 BC, and lathes, known to have existed in multiple regions of Europe since at least 1000 to 500 BC. But it was not until the later Middle Ages and the Age of Enlightenment that the modern concept of a machine tool—a class of machines used as tools in the making of metal parts, and incorporating machine-guided toolpath—began to evolve. Clockmakers of the Middle Ages and renaissance men such as Leonardo da Vinci helped expand humans' technological milieu toward the preconditions for industrial machine tools. During the 18th and 19th centuries, and even in many cases in the 20th, the builders of machine tools tended to be the same people who would then use them to produce the end products (manufactured goods). However, from these roots also evolved an industry of machine tool builders as we define them today, meaning people who specialize in building machine tools for sale to others.

Historians of machine tools often focus on a handful of major industries that most spurred machine tool development. In order of historical emergence, they have been firearms (small arms and artillery); clocks; textile machinery; steam engines (stationary, marine, rail, and otherwise) (the story of how Watt's need for an accurate cylinder spurred Boulton's boring machine is discussed by Roe); sewing machines; bicycles; automobiles; and aircraft. Others could be included in this list as well, but they tend to be connected with the root causes already listed. For example, rolling-element bearings are an industry of themselves, but this industry's main drivers of development were the vehicles already listed—trains, bicycles, automobiles, and aircraft; and other industries, such as tractors, farm implements, and tanks, borrowed heavily from those same parent industries.

Machine tools filled a need created by textile machinery during the Industrial Revolution in England in the middle to late 1700s. Until that time machinery was made mostly from wood, often including gearing and shafts. The increase in mechanization required more metal parts, which were usually made of cast iron or wrought iron. Cast iron could be cast in molds for larger parts, such as engine cylinders and gears, but was difficult to work with a file and could not be hammered. Red hot wrought iron could be hammered into shapes. Room temperature wrought iron was worked with a file and chisel and could be made into gears and other complex parts; however, hand working lacked precision and was a slow and expensive process.

James Watt was unable to have an accurately bored cylinder for his first steam engine, trying for several years until John Wilkinson invented a suitable boring machine in 1774, boring Boulton & Watt's first commercial engine in 1776.

The advance in the accuracy of machine tools can be traced to Henry Maudslay and refined by Joseph Whitworth. That Maudslay had established the manufacture and use of master plane gages in his shop (Maudslay & Field) located on Westminster Road south of the Thames River in London about 1809, was attested to by James Nasmyth who was employed by Maudslay in 1829 and Nasmyth documented their use in his autobiography.

The process by which the master plane gages were produced dates back to antiquity but was refined to an unprecedented degree in the Maudslay shop. The process begins with three plates each given an identification (ex., 1,2 and 3). The first step is to rub plates 1 and 2 together with a marking medium (called bluing today) revealing the high spots which would be removed by hand scraping with a steel scraper, until no irregularities were visible. This would not produce absolutely true plane surfaces but a "ball and socket" fit, as this mechanical fit, like two perfect planes, can slide over each other and reveal no high spots. Next, plate number 3 would be compared and scraped to conform to plate number 1. In this manner plates number 2 and 3 would be identical. Next plates number 2 and 3 would be checked against each other to determine what condition existed, either both plates were "balls" or "sockets". These would then be scraped until no high spots existed and then compared to plate number 1. After repeating this process, comparing and scraping the three plates together, they would automatically generate exact true plane surfaces accurate to within millionths of an inch.

The traditional method of producing the surface gages used an abrasive powder rubbed between the plates to remove the high spots, but it was Whitworth who contributed the refinement of replacing the grinding with hand scraping. Sometime after 1825 Whitworth went to work for Maudslay and it was there that Whitworth perfected the hand scraping of master surface plane gages. In his paper presented to the British Association for the Advancement of Science at Glasgow in 1840, Whitworth pointed out the inherent inaccuracy of grinding due to no control and thus unequal distribution of the abrasive material between the plates which would produce uneven removal of material from the plates.

With the creation of master plane gages of such high accuracy, all critical components of machine tools (i.e., guiding surfaces such as machine ways) could then be compared against them and scraped to the desired accuracy. The first machine tools offered for sale (i.e., commercially available) were constructed by Matthew Murray in England around 1800. Others, such as Henry Maudslay, James Nasmyth, and Joseph Whitworth, soon followed the path of expanding their entrepreneurship from manufactured end products and millwright work into the realm of building machine tools for sale.

Important early machine tools included the slide rest lathe, screw-cutting lathe, turret lathe, milling machine, pattern tracing lathe, shaper, and metal planer, which were all in use before 1840. With these machine tools the decades-old objective of producing interchangeable parts was finally realized. An important early example of something now taken for granted was the standardization of screw fasteners such as nuts and bolts. Before about the beginning of the 19th century, these were used in pairs, and even screws of the same machine were generally not interchangeable. Methods were developed to cut screw thread to a greater precision than that of the feed screw in the lathe being used. This led to the bar length standards of the 19th and early 20th centuries.

American production of machine tools was a critical factor in the Allies' victory in World War II. Production of machine tools tripled in the United States in the war. No war was more industrialized than World War II, and it has been written that the war was won as much by machine shops as by machine guns.

The production of machine tools is concentrated in about 10 countries worldwide: China, Japan, Germany, Italy, South Korea, Taiwan, Switzerland, USA, Austria, Spain and a few others. Machine tool innovation continues in several public and private research centers worldwide.

Drive Power Sources

"all the turning of the iron for the cotton machinery built by Mr. Slater was done with hand chisels or tools in lathes turned by cranks with hand power". David Wilkinson

Machine tools can be powered from a variety of sources. Human and animal power (via cranks, treadles, treadmills, or treadwheels) were used in the past, as was water power (via water wheel); however, following the development of high-pressure steam engines in the mid 19th century, factories increasingly used steam power. Factories also used hydraulic and pneumatic power. Many small workshops continued to use water, human and animal power until electrification after 1900.

Today most machine tools are powered by electricity; however, hydraulic and pneumatic power are sometimes used, but this is uncommon.

Automatic Control

Machine tools can be operated manually, or under automatic control. Early machines used flywheels to stabilize their motion and had complex systems of gears and levers to control the machine and the piece being worked on. Soon after World War II, the numerical control (NC) machine was developed. NC machines used a series of numbers punched on paper tape or punched cards to control their motion. In the 1960s, computers were added to give even more flexibility to the process. Such machines became known as computerized numerical control (CNC) machines. NC and CNC machines could precisely repeat sequences over and over, and could produce much more complex pieces than even the most skilled tool operators.

Before long, the machines could automatically change the specific cutting and shaping tools that were being used. For example, a drill machine might contain a magazine with a variety of drill bits for producing holes of various sizes. Previously, either machine operators would usually have to manually change the bit or move the work piece to another station to perform these different operations. The next logical step was to combine several different machine tools together, all under computer control. These are known as machining centers, and have dramatically changed the way parts are made.

From the simplest to the most complex, most machine tools are capable of at least partial self-replication, and produce machine parts as their primary function.

Examples

Examples of machine tools are:

- Broaching machine

- Drill press

- Gear shaper

- Hobbing machine

- Hone

- Lathe

- Screw machines

- Milling machine

- Shear (sheet metal)

- Shaper

- Saws

- Planer

- Stewart platform mills

- Grinding machines

- Multitasking machines (MTMs)—CNC machine tools with many axes that combine turning, milling, grinding, and material handling into one highly automated machine tool

When fabricating or shaping parts, several techniques are used to remove unwanted metal. Among these are:

- Electrical discharge machining

- Grinding (abrasive cutting)

- Multiple edge cutting tools

- Single edge cutting tools

Other techniques are used to *add* desired material. Devices that fabricate components by selective *addition* of material are called rapid prototyping machines.

Machine Tool Manufacturing Industry

The worldwide market for machine tools was approximately $81 billion in production in 2014 according to a survey by market research firm Gardner Research. The largest producer of machine tools was China with $23.8 billion of production followed by Germany and Japan at neck and neck with $12.9 billion and $12.8 billion respectively. South Korea and Italy rounded out the top 5 producers with revenue of $5.6 billion and $5 billion respectively.

References

- Thomson, Ross (2009), Structures of Change in the Mechanical Age: Technological Invention in the United States 1790-1865, Baltimore, MD: The Johns Hopkins University Press, ISBN 978-0-8018-9141-0.

- Herman, Arthur. Freedom's Forge: How American Business Produced Victory in World War II, pp. 87, 112, 121, 146-50, 161, Random House, New York, NY. ISBN 978-1-4000-6964-4.

- Parker, Dana T. Building Victory: Aircraft Manufacturing in the Los Angeles Area in World War II, pp. 5, 7-8, Cypress, CA, 2013. ISBN 978-0-9897906-0-4.

- Zelinski, Peter (2014-02-21), "The capacity to build 3D metal forms is a retrofittable option for subtractive CNC machine tools", Modern Machine Shop Additive Manufacturing supplement.

- Zelinski, Peter (2013-11-08), "Hybrid machine combines milling and additive manufacturing", Modern Machine Shop.

Types of Machine Tool

Broaching is the process that is uses broach to remove material. The two types of broaches are linear and rotary. Some of the types of machine tools discussed within this section are drills, power tools, hand tools, shapers and four-slides. The topics discussed in the section are of great importance to broaden the existing knowledge on machine tools.

Broaching (Metalworking)

Broaching is a machining process that uses a toothed tool, called a broach, to remove material. There are two main types of broaching: *linear* and *rotary*. In linear broaching, which is the more common process, the broach is run linearly against a surface of the workpiece to effect the cut. Linear broaches are used in a broaching machine, which is also sometimes shortened to *broach*. In rotary broaching, the broach is rotated and pressed into the workpiece to cut an axis symmetric shape. A rotary broach is used in a lathe or screw machine. In both processes the cut is performed in one pass of the broach, which makes it very efficient.

A push style $\frac{5}{16}$ inch (8 mm) keyway broach; note how the teeth are larger on the left end.

Broaching is used when precision machining is required, especially for odd shapes. Commonly machined surfaces include circular and non-circular holes, splines, keyways, and flat surfaces. Typical workpieces include small to medium-sized castings, forgings, screw machine parts, and stampings. Even though broaches can be expensive, broaching is usually favored over other processes when used for high-quantity production runs.

Broaches are shaped similar to a saw, except the height of the teeth increases over the length of the tool. Moreover, the broach contains three distinct sections: one for roughing, another for semi-finishing, and the final one for finishing. Broaching is an unusual machining process because it has the feed built into the tool. The profile of the machined surface is always the inverse of the profile of the broach. The rise per tooth

(RPT), also known as the *step* or feed per tooth, determines the amount of material removed and the size of the chip. The broach can be moved relative to the workpiece or vice versa. Because all of the features are built into the broach no complex motion or skilled labor is required to use it. A broach is effectively a collection of single-point cutting tools arrayed in sequence, cutting one after the other; its cut is analogous to multiple passes of a shaper.

A broached keyway in the end of an adjustable wrench.

Process

The process depends on the type of broaching being performed. Surface broaching is very simple as either the workpiece is moved against a stationary surface broach, or the workpiece is held stationary while the broach is moved against it.

Internal broaching is more involved. The process begins by clamping the workpiece into a special holding fixture, called a *workholder*, which mounts in the broaching machine. The broaching machine *elevator*, which is the part of the machine that moves the broach above the workholder, then lowers the broach through the workpiece. Once through, the broaching machine's *puller*, essentially a hook, grabs the *pilot* of the broach. The elevator then releases the top of the pilot and the puller pulls the broach through the workpiece completely. The workpiece is then removed from the machine and the broach is raised back up to reengage with the elevator. The broach usually only moves linearly, but sometimes it is also rotated to create a spiral spline or gun-barrel rifling.

Cutting fluids are used for three reasons;

1. to cool the workpiece and broach

2. to lubricate cutting surfaces

3. to flush the chips from the teeth.

Fortified petroleum cutting fluids are the most common. However, heavy-duty water-soluble cutting fluids are being used because of their superior cooling, cleanliness, and non-flammability.

Usage

Broaching was originally developed for machining internal keyways. However, it was soon discovered that broaching is very useful for machining other surfaces and shapes for high volume workpieces. Because each broach is specialized to cut just one shape either the broach must be specially designed for the geometry of the workpiece or the workpiece must be designed around a standard broach geometry. A customized broach is usually only viable with high volume workpieces, because the broach can cost US$15,000 to US$30,000 to produce.

An example of a broached workpiece. Here the broaching profile is a spline.

Broaching speeds vary from 20 to 120 surface feet per minute (SFPM). This results in a complete cycle time of 5 to 30 seconds. Most of the time is consumed by the return stroke, broach handling, and workpiece loading and unloading.

The only limitations on broaching are that there are no obstructions over the length of the surface to be machined, the geometry to be cut does not have curves in multiple planes, and that the workpiece is strong enough to withstand the forces involved. Specifically for internal broaching a hole must first exist in the workpiece so the broach can enter. Also, there are limits on the size of internal cuts. Common internal holes can range from 0.125 to 6 in (3.2 to 152.4 mm) in diameter but it is possible to achieve a range of 0.05 to 13 in (1.3 to 330.2 mm). Surface broaches' range is usually 0.075 to 10 in (1.9 to 254.0 mm), although the feasible range is 0.02 to 20 in (0.51 to 508.00 mm).

Tolerances are usually ±0.002 in (±0.05 mm), but in precise applications a tolerance of ±0.0005 in (±0.01 mm) can be held. Surface finishes are usually between 16 and 63 microinches (µin), but can range from 8 to 125 µin. There may be minimal burrs on the exit side of the cut.

Broaching works best on softer materials, such as brass, bronze, copper alloys, aluminium, graphite, hard rubbers, wood, composites, and plastic. However, it still has a good machinability rating on mild steels and free machining steels. When broaching, the machinability rating is closely related to the hardness of the material. For steels the ideal hardness range is between 16 and 24 Rockwell C (HRC); a hardness greater than HRC 35 will dull the broach quickly. Broaching is more difficult on harder materials, stainless steel and titanium, but is still possible.

Types

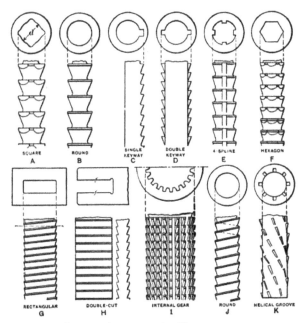

Broaches can be categorized by many means:

- Use: internal, or surface

- Purpose: single, or combination

- Motion: push, pull, or stationary

- Construction: solid, built-up, hollow or shell

- Function: roughing, sizing, or burnishing

If the broach is large enough the costs can be reduced by using a *built-up* or *modular* construction. This involves producing the broach in pieces and assembling it. If any portion wears out only that section has to be replaced, instead of the entire broach.

Most broaches are made from high speed steel (HSS) or an alloy steel; TiN coatings are common on HSS to prolong life. Except when broaching cast iron, tungsten carbide is rarely used as a tooth material because the cutting edge will crack on the first pass.

Surface Broaches

The *slab broach* is the simplest surface broach. It is a general purpose tool for cutting flat surfaces.

Slot broaches (G & H) are for cutting slots of various dimensions at high production rates. Slot broaching is much quicker than milling when more than one slot needs to be machined, because multiple broaches can be run through the part at the same time on the same broaching machine.

Contour broaches are designed to cut concave, convex, cam, contoured, and irregular shaped surfaces.

Pot broaches are cut the inverse of an internal broach; they cut the outside diameter of a cylindrical workpiece. They are named after the pot looking fixture in which the broaches are mounted; the fixture is often referred to as a "pot". The pot is designed to hold multiple broaching tools concentrically over its entire length. The broach is held stationary while the workpiece is pushed or pulled through it. This has replaced hobbing for some involute gears and cutting external splines and slots.

Straddle broaches use two slab broaches to cut parallel surfaces on opposite sides of a workpiece in one pass. This type of broaching holds closer tolerances than if the two cuts were done independently. It is named after the fact that the broaches "straddle" the workpiece on multiple sides.

Internal Broaches

Solid broaches are the most common type; they are made from one solid piece of material. For broaches that wear out quickly *shell* broaches are used; these broaches are similar to a solid broach, except there is a hole through the center where it mounts on an arbor. Shell broaches cost more initially, but save the cost overall if the broach must be replaced often because the pilots are on the mandrel and do not have to be reproduced with each replacement.

A modular broach

Modular broaches are commonly used for large internal broaching applications. They are similar to shell broaches in that they are a multi-piece construction. This design is used because it is cheaper to build and resharpen and is more flexible than a solid design.

A common type of internal broach is the *keyway* broach (C & D). It uses a special fixture called a *horn* to support the broach and properly locate the part with relation to the broach.

A *concentricity broach* is a special type of spline cutting broach which cuts both the minor diameter and the spline form to ensure precise concentricity.

The *cut-and-recut broach* is used to cut thin-walled workpieces. Thin-walled workpieces have a tendency to expand during cutting and then shrink afterward. This broach overcomes that problem by first broaching with the standard roughing teeth, followed by a "breathing" section, which serves as a pilot as the workpiece shrinks. The teeth after the "breathing" section then include roughing, semi-finishing, and finishing teeth.

An internal broach for cutting splines

The finishing teeth

The semi-finishing teeth

The roughing teeth

The front pilot

The slot in the tip of the broach where the broaching machine latches on to the broach to pull it through the workpiece

Design

For defining the geometry of a broach an internal type is shown below. Note that the geometries of other broaches are similar.

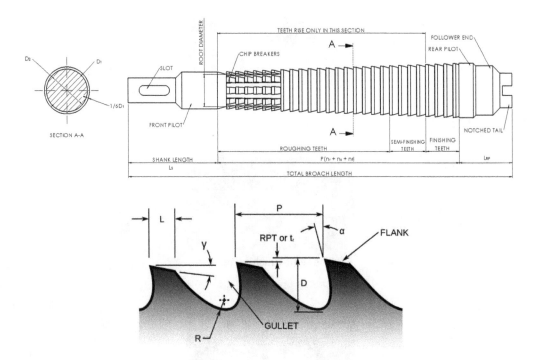

where:

- P = pitch

- RPT = rise per tooth

- n_r = number of roughing teeth

- n_s = number of semi-finishing teeth

- n_f = number of finishing teeth

- t_r = RPT for the roughing teeth

- t_s = RPT for the semi-finishing teeth

- t_f = RPT for the finishing teeth

- L_s = Shank length

- L_{RP} = Rear pilot length

- D_1 = Diameter of the tooth tip

- D_2 = Diameter of the tooth root

- D = Depth of a tooth (0.4P)

- L = Land (behind the cutting edge) (0.25P)

- R = Radius of the gullet (0.25P)

- α = Hook angle or rake angle

- γ = Back-off angle or clearance angle

- L_w = Length of the workpiece

A progressive surface broach

The most important characteristic of a broach is the rise per tooth (RPT), which is how much material is removed by each tooth. The RPT varies for each section of the broach, which are the roughing section (t_r), semi-finishing section (t_s), and finishing section (t_f). The roughing teeth remove most of the material so the number of roughing teeth required dictates how long the broach is. The semi-finishing teeth provide surface finish and the finishing teeth provide the final finishing. The finishing section's RPT (t_f) is usually zero so that as the first finishing teeth wear the later ones continue the sizing function. For free-machining steels the RPT ranges from 0.006 to 0.001 in (0.152 to 0.025 mm). For surface broaching the RPT is usually between 0.003 to 0.006 in (0.076 to 0.152 mm) and for diameter broaching is usually between 0.0012 to 0.0025 in (0.030 to 0.064 mm). The exact value depends on many factors. If the cut is too big it will impart too much stress into the teeth and the workpiece; if the cut is too small the teeth rub instead of cutting. One way to increase the RPT while keeping the stresses down is with *chip breakers*. They are notches in the teeth designed to break the chip and decrease the overall amount of material being removed by any given tooth. For broaching to be effective, the workpiece should have 0.020 to 0.025 in (0.51 to 0.64 mm) more material than the final dimension of the cut.

The *hook* (a) angle is a parameter of the material being cut. For steel, it is between 15 and 20° and for cast iron it is between 6 and 8°. The *back-off* (γ) provides clearance for the teeth so that they don't rub on the workpiece; it is usually between 1 and 3°.

When radially broaching workpieces that require a deep cut per tooth, such as forgings or castings, a *rotor-cut* or *jump-cut* design can be used; these broaches are also known as *free egress* or *nibbling* broaches. In this design the RPT is designated to two or three rows of teeth. For the broach to work the first tooth of that cluster has a wide notch, or undercut, and then the next tooth has a smaller notch (in a three tooth design) and the final tooth has no notch. This allows for a deep cut while keeping stresses, forces, and power requirements low.

There are two different options for achieving the same goal when broaching a flat surface. The first is similar to the rotor-cut design, which is known as a *double-cut* design. Here four teeth in a row have the same RPT, but each progressive tooth takes only a portion of the cut due to notches in the teeth. The other option is known as a *progressive* broach, which completely machines the center of the workpiece and then the rest of the broach machines outward from there. All of these designs require a broach that is longer than if a standard design were used.

For some circular broaches, *burnishing teeth* are provided instead of finishing teeth. They are not really teeth, as they are just rounded discs that are 0.001 to 0.003 in (0.025 to 0.076 mm) oversized. This results in burnishing the hole to the proper size. This is primarily used on non-ferrous and cast iron workpieces.

The pitch defines the tooth construction, strength, and number of teeth in contact with the workpiece. The pitch is usually calculated from workpiece length, so that the broach can be designed to have at least two teeth in contact with the workpiece at any time; the pitch remains constant for all teeth of the broach. One way to calculate the pitch is:

$$P \cong 0.35\,L\,w$$

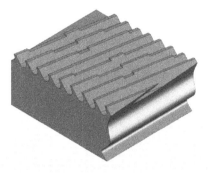

Example of a double-cut surface broach

Top view of a double-cut surface broach

Side view of a double-cut surface broach
Broaching machines

The hydraulic cylinder of a horizontal broaching machine.

Broaching machines are relatively simple as they only have to move the broach in a linear motion at a predetermined speed and provide a means for handling the broach automatically. Most machines are hydraulic, but a few specialty machines are mechanically driven. The machines are distinguished by whether their motion is horizontal or vertical. The choice of machine is primarily dictated by the stroke required. Vertical broaching machines rarely have a stroke longer than 60 in (1.5 m).

Vertical broaching machines can be designed for push broaching, pull-down broaching, pull-up broaching, or surface broaching. Push broaching machines are similar to an arbor press with a guided ram; typical capacities are 5 to 50 tons. The two ram pull-down machine is the most common type of broaching machine. This style machine has the rams under the table. Pull-up machines have the ram above the table; they usually have more than one ram. Most surface broaching is done on a vertical machine.

Horizontal broaching machines are designed for pull broaching, surface broaching, continuous broaching, and rotary broaching. Pull style machines are basically vertical machines laid on the side with a longer stroke. Surface style machines hold the broach stationary while the workpieces are clamped into fixtures that are mounted on a conveyor system. Continuous style machines are similar to the surface style machines except adapted for internal broaching.

Horizontal machines used to be much more common than vertical machines; however, today they represent just 10% of all broaching machines purchased. Vertical machines are more popular because they take up less space.

Broaching is often impossible without the specific broaching or keyway machines unless you have a system that can be used in conjunction with a modern machining centre or driven tooling lathe; these extra bits of equipment open up the possibility of producing keyways, splines and torx through one-hit machining.

Rotary Broaching

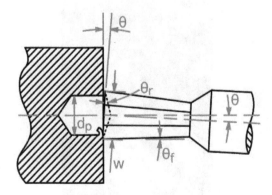

Schematic of a rotary broach starting a cut.

θ	Off-axis (wobble) angle
θ_r	Rake
θ_f	Front relief
d_p	Pilot diameter
w	Width across corners (AC)

A somewhat different design of cutting tool that can achieve the irregular hole or outer profile of a broach is called a *rotary broach* or *wobble broach*. One of the biggest advantages to this type of broaching is that it does not require a broaching machine, but instead is used on lathes, milling machines, screw machines or Swiss lathes.

Rotary broaching requires two tooling components: a tool holder and a broach. The leading (cutting) edge of the broach has a contour matching the desired final shape. The broach is mounted in a special tool holder that allows it to freely rotate. The tool holder is special because it holds the tool so that its axis of rotation is inclined slightly to the axis of rotation of the work. A typical value for this misalignment is 1°. This angle is what produces a rotating edge for the broach to cut the workpiece. Either the workpiece or the tool holder is rotated. If the tool holder is rotated, the misalignment causes the broach to appear as though it is "wobbling", which is the origin of the term *wobble broach*.

For internal broaching the sides of the broach are drafted inward so it becomes thinner; for external broaching the sides are drafted outward, to make the pocket bigger. This draft keeps the broach from jamming; the draft must be larger than the angle of misalignment. If the work piece rotates, the broach is pressed against it, is driven by it, and rotates synchronously with it. If the tool holder rotates, the broach is pressed against the workpiece, but is driven by the tool holder.

Ideally the tool advances at the same rate that it cuts. The ideal rate of cut is defined as:

> Rate of cut [inches per rotation (IPR)] = (diameter of tool [inches]) × sin(Angle of misalignment [degrees])

If it advances much faster, then the tool becomes choked; conversely, if it advances much slower, then an interrupted or zig-zag cut occurs. In practice the rate of cut is slightly less than the ideal rate so that the load is released on the non-cutting edge of the tool.

There is some spiraling of the tool as it cuts, so the form at the bottom of the workpiece may be rotated with respect to the form at the top of the hole or profile. Spiraling may be undesirable because it binds the body of the tool and prevents it from cutting sharply. One solution to this is to reverse the rotation in mid cut, causing the tool to spiral in the opposite direction. If reversing the machine is not practical, then interrupting the cut is another possible solution.

In general, a rotary broach will not cut as accurately as a push or pull broach. However, the ability to use this type of cutting tool on common machine tools is highly advantageous.

History

The concept of broaching can be traced back to the early 1850s, with the first applications used for cutting keyways in pulleys and gears. After World War I, broaching was used to rifle gun barrels. In the 1920s and 30s the tolerances were tightened and the cost reduced thanks to advances in form grinding and broaching machines.

Drill

A drill is a tool fitted with a cutting tool attachment or driving tool attachment, usually a drill bit or driver bit, used for boring holes in various materials or fastening various materials together with the use of fasteners. The attachment is gripped by a chuck at one end of the drill and rotated while pressed against the target material. The tip, and sometimes edges, of the cutting tool does the work of cutting into the target material. This may be slicing off thin shavings (twist drills or auger bits), grinding off small particles (oil drilling), crushing and removing pieces of the workpiece (SDS masonry drill), countersinking, counterboring, or other operations.

A young boy using a cordless drill.

Drills are commonly used in woodworking, metalworking, construction and do-it-yourself projects. Specially designed drills are also used in medicine, space missions and other applications. Drills are available with a wide variety of performance characteristics, such as power and capacity.

History

Around 35,000 BC,(37016 years before now) Homo sapiens discovered the benefits of the application of rotary tools. This would have rudimentarily consisted of a pointed rock being spun between the hands to bore a hole through another material. This led to the hand drill, a smooth stick, that was sometimes attached to flint point, and was rubbed between the palms. This was used by many ancient civilizations around the world including the Mayans. The earliest perforated artifacts, such as bone, ivory, shells and antlers found, are from the Upper Paleolithic era.

A wooden drill handle and other carpentry tools found on board the 16th century carrack *Mary Rose.*

Bow drill (strap-drills) are the first machine drills, as they convert a back-and forth motion to a rotary motion, and they can be traced back to around 10,000 years ago. It was discovered that tying a cord around a stick, and then attaching the ends of the string to the ends of a stick(a bow), allowed a user to drill quicker and more efficiently. Mainly used to create fire, bow-drills were also used in ancient woodwork, stonework and dentistry. Archeologist discovered a Neolithic grave yard in Mehrgrath, Pakistan dating from the time of the Harappans, around 7,500–9,000 years ago, containing 9 adult bodies with a total of 11 teeth that had been drilled. There are hieroglyphs depicting Egyptian carpenters and bead makers in a tomb at Thebes using bow-drills. The earliest evidence of these tools being used in Egypt dates back to around 2500 BCE. The usage of bow-drills was widely spread through Europe, Africa, Asia and North America, during ancient times and is still used today. Over the years many slight variations of bow and strap drills have developed for the various uses of either boring through materials or lighting fires.

The core drill was developed in ancient Egypt by 3000 BC(5016 years before now). The pump drill was invented during Roman times. It consists of a vertical spindle aligned by a piece of horizontal wood and a flywheel to maintain accuracy and momentum.

The hollow-borer tip, first used around the 13th century, consisted of a stick with a tubular shaped piece of metal on the end, such as copper. This allowed a hole to be drilled while only actually grinding the outer section of it. This completely separates the inner stone or wood from the rest, allowing the drill to pulverize less material to create a similarly sized hole.

While the pump-drill and the bow-drill were used in Western Civilization to bore smaller holes for a larger part of human history, the Auger was used to drill larger holes starting sometime between Roman and Medieval ages. The auger allowed for more torque for larger holes. It is uncertain when the Brace and Bit was invented; however, the earliest picture found so far dates from the 15th century. It is a type of hand crank drill that consists of two parts as seen in the picture. The brace, on the upper half, is where the user holds and turns it and on the lower part is the bit. The bit is interchangeable as bits wear down. The auger uses a rotating helical screw similar to the Archimedean screw-shaped bit that is common today. The gimlet is also worth mentioning as it is a scaled down version of an auger.

In the East, churn drills were invented as early as 221 BC during the Chinese Qin Dynasty, capable of reaching a depth of 1500 m. Churn drills in ancient China were built of wood and labor-intensive, but were able to go through solid rock. The churn drill appears in Europe during the 12th century. In 1835 Isaac Singer is reported to have built a steam powered churn drill based off the method the Chinese used. Also worth briefly discussing are the early drill presses; they were machine tools that derived from bow-drills but were powered by windmills and water wheels. Drill presses consisted of the powered drills that could be raised or lowered into a material, allowing for less force by the user.

The next great advancement in drilling technology, the electric motor, led to the invention of the electric drill. It is credited to Arthur James Arnot and William Blanch Brain of Melbourne, Australia who patented the electric drill in 1889. In 1895, the first portable handheld drill was created by brothers Wilhem & Carl Fein of Stuttgart, Germany. In 1917 the first trigger-switch, pistol-grip portable drill was patented by Black & Decker. This was the start of the modern drill era. Over the last century the electric drill has been created in a variety of types and multiple sizes for an assortment of specific uses.

Types

There are many types of drills: some are powered manually, others use electricity (electric drill) or compressed air (*pneumatic drill*) as the motive power, and a minority are driven by an internal combustion engine (for example, earth drilling augers). Drills with a percussive action (hammer drills) are mostly used in hard materials such as masonry (brick, concrete and stone) or rock. Drilling rigs are used to bore holes in the earth to obtain water or oil. Oil wells, water wells, or holes for geothermal heating are created with large drilling rigs. Some types of hand-held drills are also used to drive screws and other fasteners. Some small appliances that have no motor of their own may be drill-powered, such as small pumps, grinders, etc.

Carpenter using a crank-powered brace to drill a hole

Hand Drills

A variety of hand-powered drills have been employed over the centuries. Here are a few, starting with approximately the oldest:

- Bow drill

- Brace and bit

- Gimlet

- Hand drill, also known as an "eggbeater" drill, or (especially in the UK) a wheel brace

- Breast drill, similar to an "eggbeater" drill, it has a flat chest piece instead of a handle

- Push drill, a tool using a spiral ratchet mechanism

- Pin chuck, a small hand-held jeweler's drill

An old hand drill or "eggbeater" drill. The hollow wooden handle, with screw-on cap, is used to store drill bits

manually operated tool for making holes.

Pistol-grip (Corded) Drill

Drills with pistol grips are the most common type in use today, and are available in a huge variety of subtypes. A less common type is the right-angle drill, a special tool used by tradesmen such as plumbers and electricians. The motor used in corded drills is often a universal motor due to its high power to weight ratio.

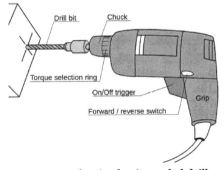

Anatomy of a pistol-grip corded drill.

For much of the 20th century, many attachments could commonly be purchased to convert corded electric hand drills into a range of other power tools, such as orbital sanders and power saws, more cheaply than purchasing conventional, self-contained versions of those tools (the greatest saving being the lack of an additional electric motor for each device). As the prices of power tools and suitable electric motors have fallen, however, such attachments have become much less common. A similar practice is currently employed for cordless tools where the battery, the most expensive component, is shared between various motorised devices, as opposed to a single electric motor being shared between mechanical attachments.

Drills can also be used at an angle to join two boards.

Hammer Drill

The hammer drill is similar to a standard electric drill, with the exception that it is provided with a hammer action for drilling masonry. The hammer action may be engaged or disengaged as required. Most electric hammer drills are rated (input power) at between 600 and 1100 watts. The efficiency is usually 50-60% i.e. 1000 watts of input is converted into 500-600 watts of output (rotation of the drill and hammering action).

The hammer action is provided by two cam plates that make the chuck rapidly pulse forward and backward as the drill spins on its axis. This pulsing (hammering) action is measured in Blows Per Minute (BPM) with 10,000 or more BPMs being common. Because the combined mass of the chuck and bit is comparable to that of the body of the drill, the energy transfer is inefficient and can sometimes make it difficult for larger bits to penetrate harder materials such as poured concrete. The operator experiences considerable vibration, and the cams are generally made from hardened steel to avoid them wearing out quickly. In practice, drills are restricted to standard masonry bits up to 13 mm (1/2 inch) in diameter. A typical application for a hammer drill is installing electrical boxes, conduit straps or shelves in concrete.

In contrast to the cam-type hammer drill, a rotary/pneumatic hammer drill accelerates only the bit. This is accomplished through a piston design, rather than a spinning cam. Rotary hammers have much less vibration and penetrate most building materials. They can also be used as "drill only" or as "hammer only" which extends their usefulness for tasks such as chipping brick or concrete. Hole drilling progress is greatly superior to cam-type hammer drills, and these drills are generally used for holes of 19 mm (3/4 inch) or greater in size. A typical application for a rotary hammer drill is boring large holes for lag bolts in foundations, or installing large lead anchors in concrete for handrails or benches.

A standard hammer drill accepts 6 mm (1/4 inch) and 13 mm (1/2 inch) drill bits, while a rotary hammer uses SDS or Spline Shank bits. These heavy bits are adept at pulverising the masonry and drill into this hard material with relative ease.

However, there is a big difference in cost. In the UK a cam hammer typically costs £12 or more, while a rotary/pneumatic costs £35 or more. In the US a typical hammer drill costs between $70 and $120, and a rotary hammer between $150 and $500 (depending on bit size). For DIY use or to drill holes less than 13 mm (1/2 inch) in size, the hammer drill is most commonly used.

Magnetic Drilling Machine

A Magnetic Drilling Machine is a machine tool used for drilling holes in variety of metals. A magnetic drilling machine is a portable drilling machine with a base or a permanent magnet base. The magnetic drilling machine drills holes with the help of cutting tools like annular cutters (broach cutters) or with twist drill bits. There are various types of magnetic drilling machines depending on their operations and specialisations like magnetic drilling cum tapping machines, cordless magnetic drilling machines, pneumatic magnetic drilling machines, compact horizontal magnetic drilling machines, automatic feed magnetic drilling machines, cross table base magnetic drilling machines etc.

A lightweight magnetic drilling machine MABasic 200 manufactured by BDS Maschinen GmbH, Germany. The image shows the magnetic drilling machine making holes on H-beam with the help of HSS annular cutter at a workshop.

A Magnetic Drilling Machine is also popularly known as a magnetic core drilling machine, magnetic base drilling machine, mag base drilling machine, mag base, magnetic broach cutter machine, magnetic drill ormag drill.

A portable magnetic drilling machine is the best solution for drilling large and heavy workpieces which are difficult to move or bring to the stationary conventional drilling machine for making holes, where as the magnetic drilling machines are so portable and lightweight that the machines can be taken to the workpiece to be drilled. The magnetic drilling machines have popularly changed the method of bringing workpiece to the drilling machine by bringing the machine to the workpiece.

Rotary Hammer Drill

The rotary hammer drill (also known as a rotary hammer, roto hammer drill or masonry drill) combines a primary dedicated hammer mechanism with a separate rotation mechanism, and is used for more substantial material such as masonry or concrete. Generally, standard chucks and drills are inadequate and chucks such as SDS and carbide drills that have been designed to withstand the percussive forces are used. Some styles of this tool are intended for masonry drilling only and the hammer action cannot be disengaged. Other styles allow the drill to be used without the hammer action for normal drilling, or hammering to be used without rotation for chiselling. In 1813 Richard Trevithick designed a steam-driven rotary drill, also the first drill to be powered by steam.

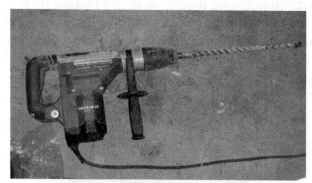

A rotary hammer drill used in construction

Cordless Drills

A cordless drill is an electric drill which uses rechargeable batteries. These drills are available with similar features to an AC mains-powered drill. They are available in the hammer drill configuration and most have a clutch, which aids in driving screws into various substrates while not damaging them. Also available are right angle drills, which allow a worker to drive screws in a tight space. While 21st century battery innovations allow significantly more drilling, large diameter holes (typically 12–25 mm (0.5–1.0 in) or larger) may drain current cordless drills quickly.

For continuous use, a worker will have one or more spare battery packs charging while drilling, and quickly swap them instead of having to wait an hour or more for recharging, although there are now Rapid Charge Batteries that can charge in 10–15 minutes.

Cordless drill

Early cordless drills used interchangeable 7.2 V battery packs. Over the years battery voltages have increased, with 18 V drills being most common, but higher voltages are available, such as 24 V, 28 V, and 36 V. This allows these tools to produce as much torque as some corded drills.

Common battery types of are nickel-cadmium (NiCd) batteries and lithium-ion batteries, with each holding about half the market share. NiCd batteries have been around longer, so they are less expensive (their main advantage), but have more disadvantages compared to lithium-ion batteries. NiCd disadvantages are limited life, self-discharging, environment problems upon disposal, and eventually internally short circuiting due to dendrite growth. Lithium-ion batteries are becoming more common because of their short charging time, longer life, absence of memory effect, and low weight. Instead of charging a tool for an hour to get 20 minutes of use, 20 minutes of charge can run the tool for an hour. Lithium-ion batteries also hold a charge for a significantly longer time than nickel-cadmium batteries, about two years if not used, vs. 1 to 4 months for a nickel-cadmium battery.

Drill Press

A drill press

Woman operating drill press (boring machine), boring wooden reels for winding barbed wire, 1917.

A drill press (also known as a pedestal drill, pillar drill, or bench drill) is a fixed style of drill that may be mounted on a stand or bolted to the floor or workbench. Portable models with a magnetic base grip the steel workpieces they drill. A drill press consists of a base, column (or pillar), table, spindle (or quill), and drill head, usually driven by an induction motor. The head has a set of handles (usually 3) radiating from a central hub that, when turned, move the spindle and chuck vertically, parallel to the axis of the column. The drill press is typically measured in terms of *swing*. Swing is defined as twice the *throat distance*, which is the distance from the center of the spindle to the closest edge of the pillar. For example, a 16-inch (410 mm) drill press has an 8-inch (200 mm) throat distance.

A drill press has a number of advantages over a hand-held drill:

- Less effort is required to apply the drill to the workpiece. The movement of the chuck and spindle is by a lever working on a rack and pinion, which gives the operator considerable mechanical advantage

- The table allows a vise or clamp to be used to position and restrain the work, making the operation much more secure

- The angle of the spindle is fixed relative to the table, allowing holes to be drilled accurately and consistently

- Drill presses are almost always equipped with more powerful motors compared to hand-held drills. This enables larger drill bits to be used and also speeds up drilling with smaller bits.

For most drill presses—especially those meant for woodworking or home use—speed change is achieved by manually moving a belt across a stepped pulley arrangement. Some drill presses add a third stepped pulley to increase the number of available speeds. Modern drill presses can, however, use a variable-speed motor in conjunction with the stepped-pulley system. Medium-duty drill presses such as those used in machine shop (tool room) applications are equipped with a continuously variable transmission. This mechanism is based on variable-diameter pulleys driving a wide, heavy-duty belt. This gives a wide speed range as well as the ability to change speed while the machine is running. Heavy-duty drill presses used for metalworking are usually of the gear-head type described below.

Drill presses are often used for miscellaneous workshop tasks other than drilling holes. This includes sanding, honing, and polishing. These tasks can be performed by mounting sanding drums, honing wheels and various other rotating accessories in the chuck. This can be unsafe in some cases, as the chuck arbor, which may be retained in the spindle solely by the friction of a taper fit, may dislodge during operation if the side loads are too high.

Geared Head Drill Press

A geared head drill press is a drill press in which power transmission from the motor to the spindle is achieved solely through spur gearing inside the machine's head. No friction elements (e.g., belts) of any kind are used, which assures a positive drive at all times and minimizes maintenance requirements. Gear head drills are intended for metalworking applications where the drilling forces are higher and the desired speed (RPM) is lower than that used for woodworking.

Geared head drill press. Shift levers on the head and a two speed motor control immediately in front of the quill handle select one of eight possible speeds

Levers attached to one side of the head are used to select different gear ratios to change the spindle speed, usually in conjunction with a two- or three-speed motor (this varies with the material). Most machines of this type are designed to be operated on three-phase electric power and are generally of more rugged construction than equivalently sized belt-driven units. Virtually all examples have geared racks for adjusting the table and head position on the column.

Radial arm drill press.

Geared head drill presses are commonly found in tool rooms and other commercial environments where a heavy duty machine capable of production drilling and quick setup changes is required. In most cases, the spindle is machined to accept Morse taper tooling for greater flexibility. Larger geared head drill presses are frequently fitted with power feed on the quill mechanism, with an arrangement to disengage the feed when a certain drill depth has been achieved or in the event of excessive travel. Some gear-head drill presses have the ability to perform tapping operations without the need for an external tapping attachment. This feature is commonplace on larger gear head drill presses. A clutch mechanism drives the tap into the part under power and then backs it out of the threaded hole once the proper depth is reached. Coolant systems are also common on these machines to prolong tool life under production conditions.

Radial Arm Drill Press

Radial arm drill press controls

A radial arm drill press is a large geared head drill press in which the head can be moved along an arm that radiates from the machine's column. As it is possible to swing the arm relative to the machine's base, a radial arm drill press is able to operate over a large area without having to reposition the workpiece. This saves considerable time because it is much faster to reposition the drill head than it is to unclamp, move, and then re-clamp the workpiece to the table. The size of work that can be handled may be considerable, as the arm can swing out of the way of the table, allowing an overhead crane or derrick to place a bulky workpiece on the table or base. A vise may be used with a radial arm drill press, but more often the workpiece is secured directly to the table or base, or is held in a fixture. Power spindle feed is nearly universal with these machines and coolant systems are common. Larger size machines often have power feed motors for elevating or moving the arm. The biggest radial arm drill presses are able to drill holes as large as four inches (101.6 millimeters) diameter in solid steel or cast iron. Radial arm drills are specified by the diameter of the column and the length of the arm. The length of the arm is usually the same as the maximum throat distance. The Radial Arm Drill pictured in this article is a 9-inch column x 3-foot arm. The maximum throat distance of this drill would be approximately 36", giving a swing of 72" (6 feet).

Mill Drill

Mill drills are a lighter alternative to a milling machine. They combine a drill press (belt driven) with the X/Y coordinate abilities of the milling machine's table and a locking collet that ensures that the cutting tool will not fall from the spindle when lateral forces are experienced against the bit. Although they are light in construction, they have the advantages of being space-saving and versatile as well as inexpensive, being suitable for light machining that may otherwise not be affordable.

Accessories

Drills are often used simply as motors to drive a variety of applications, in much the same way that tractors with generic PTOs are used to power ploughs, mowers, trailers, etc.

Accessories available for drills include:

- Screw-driving tips of various kinds - flathead, Philips, etc. for driving screws in or out
- Water pumps
- Nibblers for cutting metal sheet
- Rotary sanding discs
- Rotary polishing discs

Drilling Capacity

Drilling capacity indicates the maximum diameter a given power drill or drill press can produce in a certain material. It is essentially a proxy for the continuous torque the machine is capable of producing. Typically a given drill will have its capacity specified for different materials, i.e., 10mm for steel, 25mm for wood, etc.

For example, the maximum recommended capacities for the DeWalt DCD790 cordless drill for specific drill bit types and materials are as follows:

Material	Drill bit type	Capacity
Wood	Auger	$\frac{7}{8}$ in (22 mm)
	Paddle	$1\frac{1}{4}$ in (32 mm)
	Twist	$\frac{1}{2}$ in (13 mm)
	Self-feed	$1\frac{3}{8}$ in (35 mm)
	Hole saw	2 in (51 mm)
Metal	Twist	$\frac{1}{2}$ in (13 mm)
	Hole saw	$1\frac{3}{8}$ in (35 mm)

Unusual Uses

- A household drill was used to save a boy's life in Australia. The boy suffered from potentially fatal bleeding within the brain after a fall from his bicycle. Lacking any surgical equipment suitable for the task, the attending doctor decided to use a household drill stored in the hospital maintenance room in order to relieve the blood pressure in the boy's brain. If this had not been done, the boy would have died in minutes. The doctor performed the procedure and was guided by a neurosurgeon over the phone. The boy was later airlifted to a larger hospital and recovered within days.

- The 1970s horror movies *The Toolbox Murders* and *The Wiener* depict deranged killers using battery powered drills to kill their victims.

- In the video game *BioShock*, certain variants of *Big Daddies* use giant arm-mounted drills to perform maintenance tasks and to kill *Splicers* that attack *Little Sisters*.

- Van Halen guitarist Eddie Van Halen has used an electric drill to create an odd effect on the strings. It has been played on songs such as Poundcake and Intruder/(Oh) Pretty Woman.

Power Tool

A power tool is a tool that is actuated by an additional power source and mechanism other than the solely manual labour used with hand tools. The most common types of power tools use electric motors. Internal combustion engines and compressed air are also commonly used. Other power sources include steam engines, direct burning of fuels and propellants, or even natural power sources like wind or moving water. Tools directly driven by animal power are not generally considered power tools.

Power tools are used in industry, in construction, in the garden, for housework tasks such as cooking, cleaning, and around the house for purposes of driving (fasteners), drilling, cutting, shaping, sanding, grinding, routing, polishing, painting, heating and more.

Power tools are classified as either stationary or portable, where portable means hand-held. Portable power tools have obvious advantages in mobility. Stationary power tools however often have advantages in speed and accuracy, and some stationary power tools can produce objects that cannot be made in any other way. Stationary power tools for metalworking are usually called *machine tools*. The term *machine tool* is not usually applied to stationary power tools for woodworking, although such usage is occasionally heard, and in some cases, such as drill presses and bench grinders, exactly the same tool is used for both woodworking and metalworking.

History

The lathe is the oldest power tool, being known to the ancient Egyptians (albeit in a hand-powered form). Early industrial revolution-era factories had batteries of power tools driven by belts from overhead shafts. The prime power source was a water wheel or (later) a steam engine.The introduction of the electric motor (and electric distribution networks) in the 1880s made possible the self-powered stationary and portable tools we know today.

Energy Sources

Currently an electric motor is the most popular choice to power stationary tools, though in the past they were powered by windmills, water wheels and steam. Some museums and hobbyists still maintain and operate stationary tools powered these older power sources. Portable electric tools may be either corded or battery-powered. Compressed air is the customary power source for nailers and paint sprayers. A few tools (called *powder-actuated tools*) are powered by explosive cartridges. Tools that run on gasoline or gasoline-oil mixes are made for outdoor use; typical examples include most chainsaws and string trimmers. Other tools like blowtorches will burn their fuel externally to generate heat. Compressed air is universally used where there is a possibility of fuel or vapor ignition - such as automotive workshops. Professional level electric tools differ from DIY or 'consumer' tools by being double insulated and not earthed - in fact they *must not* be earthed for safety reasons.

Safety

While hand-held power tools are extremely helpful, they also produce large amounts of noise and vibrations. Using power tools without hearing protection over a long period of time can put a person at risk for hearing loss. The US National Institute for Occupational Safety and Health (NIOSH) has recommended that a person should not be exposed to noise at or above 85 dB, for the sake of hearing loss prevention. Most power tools, including drills, circular saws, belt sanders, and chainsaws, operate at sound levels above the 85 dB limit, some even reaching over 100 dB. NIOSH strongly recommends wearing hearing protection while using these kinds of power tools.

Prior to the 1930s, power tools were often housed in cast metal housings. The cast metal housings were heavy, contributing to repetitive use injuries, as well as conductive - often shocking the user. As Henry Ford adapted to the manufacturing needs of World War II, he requested that A. H. Peterson, a tool manufacturer, create a lighter electric drill that was more portable for his assembly line workers. At this point, the Hole-Shooter, a drill that weighed 5 lbs. was created by A. H. Peterson. The Peterson Company eventually went bankrupt after a devastating fire and recession, but the company was auctioned off to A. F. Siebert, a former partner in the Peterson Company, in 1924 and became the Milwaukee Electric Tool Company.

In the early 30's, companies started to experiment with housings of thermoset polymer plastics. In 1956, under the influence of Dr. Hans Erich Slany, Robert Bosch GmbH was one of the first companies to introduce a power tool housing made of glass filled nylon.

List of Power Tools

Power tools include:

• Air compressor	• Lathe
• Alligator shear	• Lawn mower
• Angle grinder	• Leaf blower
• Bandsaw	• Miter saw
• Belt sander	• Multi-tool
• Biscuit joiner	• Nail gun (electric and battery as well as powder actuated)
• Ceramic tile cutter	• Needle scaler
• Chainsaw	• Pneumatic torque wrench
• Circular saw	• Powder-actuated tools
• Concrete saw	• Power wrench
• Cold saw	• Pressure washer
• Crusher	• Radial arm saw
• Diamond blade	• Random orbital sander
• Diamond tool	• Reciprocating saw
• Disc cutter	• Rotary reciprocating saw
• Disc sander	• Rotary tool
• Drill	• Rotovator
• Floor sander	• Sabre saw
• Food processor	• Sander
• Grinding machine	• Scrollsaw
• Heat gun	• Sewing machine
• Hedge trimmer	• Snow blower
• Impact driver	• Steel cut off saw
• Impact wrench	• Strimmer
• Iron	• String trimmer
• Jackhammer	• Table saw
• Jointer	• Thickness planer
• Jigsaw	• Vacuum cleaner
• Knitting machine	• Wall chaser
	• Washing machine
	• Wood router

Grinding Machine

A grinding machine, often shortened to grinder, is any of various power tools or machine tools used for grinding, which is a type of machining using an abrasive wheel as the cutting tool. Each grain of abrasive on the wheel's surface cuts a small chip from the workpiece via shear deformation.

Rotating abrasive wheel on a bench grinder.

Grinding is used to finish workpieces that must show high surface quality (e.g., low surface roughness) and high accuracy of shape and dimension. As the accuracy in dimensions in grinding is on the order of 0.000025 mm, in most applications it tends to be a finishing operation and removes comparatively little metal, about 0.25 to 0.50 mm depth. However, there are some roughing applications in which grinding removes high volumes of metal quite rapidly. Thus, grinding is a diverse field.

Pedal-powered grinding machine, Russia, 1902.

Introduction

The grinding machine consists of a bed with a fixture to guide and hold the work piece, and a power-driven grinding wheel spinning at the required speed. The speed is determined by the wheel's diameter and manufacturer's rating. The grinding head can travel across a fixed work piece, or the work piece can be moved while the grind head stays in a fixed position.

Fine control of the grinding head or table position is possible using a vernier calibrated hand wheel, or using the features of numerical controls.

Grinding machines remove material from the work piece by abrasion, which can generate substantial amounts of heat. To cool the work piece so that it does not overheat and go outside its tolerance, grinding machines incorporate a coolant. The coolant also benefits the machinist as the heat generated may cause burns. In high-precision grinding machines (most cylindrical and surface grinders), the final grinding stages are usually set up so that they remove about 200 nm (less than 1/10000 in) per pass - this generates so little heat that even with no coolant, the temperature rise is negligible.

Types

A cylindrical grinder.

A surface grinder.

These machines include the:

- Belt grinder, which is usually used as a machining method to process metals and other materials, with the aid of coated abrasives. Sanding is the machining of wood; grinding is the common name for machining metals. Belt grinding is a versatile process suitable for all kind of applications like finishing, deburring, and stock removal.

- Bench grinder, which usually has two wheels of different grain sizes for roughing and finishing operations and is secured to a workbench or floor stand. Its uses include shaping tool bits or various tools that need to be made or repaired. Bench grinders are manually operated.

- Cylindrical grinder, which includes both the types that use centers and the centerless types. A cylindrical grinder may have multiple grinding wheels. The workpiece is rotated and fed past the wheel(s) to form a cylinder. It is used to make precision rods, tubes, bearing races, bushings, and many other parts.

- Surface grinder which includes the wash grinder. A surface grinder has a "head" which is lowered to a workpiece which is moved back and forth under the grinding wheel on a table that typically has a controllable permanent magnet for use with magnetic stock but can have a vacuum chuck or other fixturing means. The most common surface grinders have a grinding wheel rotating on a horizontal axis cutting around the circumference of the grinding wheel. Rotary surface grinders, commonly known as "Blanchard" style grinders, have a grinding head which rotates the grinding wheel on a vertical axis cutting on the end face of the grinding wheel, while a table rotates the workpiece in the opposite direction underneath. This type of machine removes large amounts of material and grinds flat surfaces with noted spiral grind marks. It can also be used to make and sharpen metal stamping die sets, flat shear blades, fixture bases or any flat and parallel surfaces. Surface grinders can be manually operated or have CNC controls.

- Tool and cutter grinder and the D-bit grinder. These usually can perform the minor function of the drill bit grinder, or other specialist toolroom grinding operations.

- Jig grinder, which as the name implies, has a variety of uses when finishing jigs, dies, and fixtures. Its primary function is in the realm of grinding holes and pins. It can also be used for complex surface grinding to finish work started on a mill.

- Gear grinder, which is usually employed as the final machining process when manufacturing a high-precision gear. The primary function of these machines is to remove the remaining few thousandths of an inch of material left by other manufacturing methods (such as gashing or hobbing).

- Die grinder, which is a high-speed hand-held rotary tool with a small diameter grinding bit. They are typically air driven (using compressed air), but can be driven with a small electric motor directly or via a flexible shaft.

Pressure Washer

Home-use pressure washer powered by a small gasoline engine

A pressure washer or power washer is a high-pressure mechanical sprayer used to remove loose paint, mold, grime, dust, mud, and dirt from surfaces and objects such as
buildings, vehicles and concrete surfaces. The volume of a pressure washer is expressed
in gallons or litres per minute, often designed into the pump and not variable. The pressure, expressed in pounds per square inch, pascals, or bar (deprecated but in common
usage), is designed into the pump but can be varied by adjusting the unloader valve.
Machines that produce pressures from 750 to 30,000 psi (5 to 200 MPa) or more are
available.

A pressure washer is used to remove old paint from a boat.

The basic pressure washer consists of a motor (either electric, internal combustion,
pneumatic or hydraulic) that drives a high-pressure water pump, a high-pressure hose
and a trigger gun-style switch. Just as a garden hose nozzle is used to increase the velocity of water, a pressure washer creates high pressure and velocity. The pump cannot
draw more water from the pipe to which the washer is connected than that source can
provide: the water supply must be adequate for the machine connected to it, as water
starvation leads to cavitation damage of the pump elements.

Different types of nozzles are available for different applications. Some nozzles create
a water jet that is in a triangular plane (fan pattern), others emit a thin jet of water that
spirals around rapidly (cone pattern). Nozzles that deliver a higher flow rate lower the
output pressure. Most nozzles attach directly to the trigger gun.

Some washers, with an appropriate nozzle, allow detergent to be introduced into the
water stream, assisting in the cleaning process. Two types of chemical injectors are
available — a high-pressure injector that introduces the chemical after the water leaves
the pump (a downstream injector) and a low-pressure injector that introduces the
chemical before water enters the pump (an upstream injector). The type of injector
used is related to the type of detergent used, as there are many chemicals that will damage a pump if an upstream injector is used.

Washers are dangerous tools and should be operated with due regard to safety instructions. The water pressure near the nozzle is powerful enough to strip flesh
from bone. Particles in the water supply are ejected from the nozzle at great ve-

locities. The cleaning process can propel objects dislodged from the surface being cleaned, also at great velocities. Pressure washers have a tendency to break up tarmac if aimed directly at it, due to high-pressure water entering cracks and voids in the surface.

Most consumer washers are electric- or petrol-powered. Electric washers plug into a normal outlet, are supplied with tap water, and typically deliver pressure up to about 2,000 psi (140 bar). Petrol washers can deliver twice that pressure, but due to the hazardous nature of the engine exhaust they are unsuitable for enclosed or indoor areas. Some models can generate hot water, which can be ideal for loosening and removing oil and grease.

Origin

The hot-water high-pressure washer was invented by Alfred Kärcher in 1950, but Frank Ofeldt in the United States claimed to have invented the steam pressure washer or "high-pressure Jenny" in 1927.

Uses

At extreme high pressure, water is used in many industrial cleaning applications requiring the removal of surface layers and for dust-free cutting of some metals and concrete. For exterior applications, gas or propane powered washers provide greater mobility than electric ones, as they do not require use in proximity to an electrical outlet, but for indoor applications, electric washers produce no exhaust and are much quieter than gas or propane washers.

A pressure washer cleaning oil-clogged tarmac

The majority of pressure washers nowadays connect to an existing water supply, like a garden hose, but some models store water in an attached tank. Usually there is an on/off button that controls the water stream and many models allow you to adjust the water pressure.

High-pressure water, in combination with special chemicals, aids in the removal of graffiti, especially when the water is hot, as a quick rinser of the softened graffiti. Sometimes a pressurized mixture of air and sand, or water and sand, is used to blast off the surface of the vandalized area, etching the surface and making it extremely difficult to use high-pressure cleaning as a follow-up process. Sandblasting as graffiti removal often over cleans a surface and is capable of leaving a permanent scar on the surface.

Cleaning siding with a pressure washer

Low-pressure washing is a common technique used in today's residential house washing market. Professional pressure washing contractors have used high pressure equipment in the past to clean residential homes, however, low-pressure, high volume pressure washing equipment has been introduced as an alternative along with the use of specific cleaning solutions to achieve effective cleaning with a very low possibility of damaging the property.

Precautions

Washers can damage surfaces: water can be forced deep into bare wood and masonry, leading to an extended drying period. Such surfaces can appear dry after a short period, but still contain significant amounts of moisture that can hinder painting or sealing efforts.

Types

Washers are classified into following groups based on the type of fuel/energy they consume.

- Electric

- Diesel

- Petrol

- Gas

- Ultra high pressure

- Hydraulic high pressure

- High-pressure steam cleaner

Specialty Washers

An insulator pressure washer is a mechanical high-pressure washer designed to re-move contamination/pollution from overhead power line insulators with the power on (energized) using low conductivity water. Cleaning is necessary to prevent flash-overs (high-voltage shorts to earth across the insulators), which can damage power line equipment.

Truck Chassis Mount Insulator Washer Unit

The basic design consists of the following components:

- Power Source (a diesel engine or a power take-off from a truck chassis)

- centrifugal water pump

- stainless steel water storage tank

- high-pressure water hose

- high-pressure dead-man type water wash gun

- Electrical grounds for the wash gun and washing system

Insulator washers typically have a pump pressure of about 1000 psi and a nozzle pres-sure between 500 and 750 psi. They have a very high flow rate (about 60 gallons per minute), necessary to provide consistent and fast cleaning.

Insulator washers have several basic design layouts as follows:

- Trailer mount for use in electricity substations

- Chassis mount for mounting on a truck bed

- Aerial platform mount for mounting on trucks that have aerial platforms

- Helicopter mounted units

The washer must be able to access the towers to be washed. If the towers are off road in rough terrain, a 4x4, 6x4 or 6x6 truck chassis may be required.

The systems must have a robust design as they are used outdoors, often in adverse climates.

Angle Grinder

An angle grinder, also known as a side grinder or disc grinder, is a handheld power tool used for cutting, grinding and polishing.

Angle grinder

Battery-powered angle grinder

Angle grinders can be powered by an electric motor, petrol engine or compressed air. The motor drives a geared head at a right-angle on which is mounted an abrasive disc or a thinner cut-off disc, either of which can be replaced when worn. Angle grinders typically have an adjustable guard and a side-handle for two-handed operation. Certain angle grinders, depending on their speed range, can be used as sanders, employing a sanding disc with a backing pad or disc. The backing system is typically made of hard plastic, phenolic resin, or medium-hard rubber depending on the amount of flexibility desired.

A member of the *Technisches Hilfswerk* using an angle grinder

Uses

Angle grinders may be used for removing excess material from a piece. There are many different kinds of discs that are used for various materials and tasks, such as cut-off discs (diamond blade), abrasive grinding discs, grinding stones, sanding discs, wire brush wheels and polishing pads. The angle grinder has large bearings to counter side forces generated during cutting, unlike a power drill, where the force is axial.

Angle grinders are widely used in metalworking and construction, as well as in emergency rescues. They are commonly found in workshops, service garages and auto body repair shops. There is a large variety of angle grinders to choose from when trying to find the right one for the job. The most important factors in choosing the right grinder are the disc size and how powerful the motor is. Other factors include power source (pneumatic or electric), rpm, and arbor size. Generally disc size and power increase together. Disc size is usually measured in inches or millimetres. Common disc sizes for angle grinders in the U.S.A. include 4, 4.5, 5, 6, 7, 9 and 12 inches. Discs for pneumatic grinders also come much smaller. Pneumatic grinders are generally used for lighter duty jobs where more precision is required. This is likely because pneumatic grinders can be small and light yet remain powerful, because they do not contain heavy copper motor windings, while it is harder for an electric grinder to maintain adequate power with smaller size. Electric grinders are more commonly used for larger, heavy duty jobs. However, there are also small electric grinders and large pneumatic grinders.

Safety

A video on vibration research done on pneumatic grinders

Through a sound pressure level and vibrations study conducted by the National Institute for Occupational Safety and Health, grinders under an unloaded condition ranged from 91 to 103 dBA. In addition, angle grinders produce sparks when cutting ferrous metals. They also produce shards cutting other materials. The blades themselves may also break. This is a great hazard to the face and eyes especially, as well as other parts of the body, and as such, a full face shield and other protective clothing must be worn. Angle grinders should never be used without their guard or handle attached; they are there as a necessary precaution for safety. All work should be securely clamped or held firmly in a vice.

History

The angle grinder was invented in 1954 by German company *Ackermann + Schmitt (FLEX-Elektrowerkzeuge GmbH)* in Steinheim an der Murr. Owing to this, in German, Dutch, Slovak, Czech, Romanian and Hungarian (and possibly in other languages), an angle grinder is colloquially called a just a "flex".

Biscuit Joiner

Lamello Top biscuit joiner

A biscuit joiner (or sometimes plate joiner) is a woodworking tool used to join two pieces of wood together. A biscuit joiner uses a small circular saw blade to cut a crescent-shaped hole (called the mouth) in the opposite edges of two pieces of wood or wood composite panels. An oval-shaped, highly dried and compressed wooden biscuit (beech or particle wood) is covered with glue, or glue is applied in the slot. The biscuit is immediately placed in the slot, and the two boards are clamped together. The wet glue expands the biscuit, further improving the bond.

Lamello Top biscuit joiner with blade extended

Edges of 16mm Medium-density fibreboard with a #0 biscuit, set up to make a right angle joint.

History

The biscuit joining system was invented in 1956 in Liestal, Switzerland by Hermann Steiner. Steiner opened his carpenter's shop in 1944, and, in the middle of the 1950s, while looking for a simple means of joining the recently introduced chipboard, invented (almost by accident) the now world-famous Lamello joining system. In the succeeding years there followed further developments such as the circular saw and the first stationary biscuit (plate) joining machine in 1956 followed by the first portable biscuit joiner for Lamello grooves in 1968. In 1969 the family operation was incorporated by the name of Lamello AG. Lamello continues to manufacture very high-end biscuit joiners such as the Lamello Top 20.

Several other companies such as Porter Cable, Dewalt, and Makita also manufacture compatible biscuit joiners, including some models with interchangeable blades, enabling the user to cut both 4" and 2" biscuit slots.

Production

Biscuits are predominantly used in joining sheet goods such as plywood, particle board and medium-density fibreboard. They are sometimes used with solid wood, replacing mortise and tenon joints as they are easier to make and almost as strong. They are also used to align pieces of wood when joined edge-to-edge in making wider panels. It is important to use the same face when cutting the slots, so the boards are perfectly flush.

Biscuits are also used to align edges of workpieces, such as when forming a 90 degree angle between workpieces. The biscuit provides a quick means of getting a perfectly flush joint, while at the same time reinforcing the joint.

Typically, the machine will have an adjustable fence, so it can be set on an angle for joining mitered pieces.

Also, there are other types of specialty biscuits available, from metal connectors, used for removable panels, to hinges, making these portable machines even more flexible.

Usage

The workpieces are brought together and the user marks the location for the biscuits. Precise measurement is not required, as the biscuits are hidden when the pieces are assembled, so a quick pencil stroke that marks both pieces where they align is all that is required. The parts are separated and the machine is used to cut the slots in each piece. The machine has reference marks on the center line of the blade for easy alignment to the marks on the materials being joined.

The body of the machine with the blade is spring-loaded and in the normal position the blade is retracted. The operator aligns the machine and uses a firm pressure to push the body forward against the base plate to make the cut. The waste material is blown out of the slot on the right of the base plate.

Because the slots are slightly longer than the biscuits, it is still possible to slide the panels sideways after the joint is assembled (before the glue sets). This fact makes the biscuit joiner easy to use, because it does not require extreme accuracy or jigs to achieve perfect joints.

The depth of the cut can be altered by an adjustable stop, the smaller base can be rotated through 90 degrees and accessories are provided for altering the offset of the base to the blade (for use with thicker or thinner materials as required). Some models allow slots to be cut at angles other than 90° to the joining face, for example 45°, which greatly speeds up the assembly of things like cabinets.

Standard Biscuit Sizes

Size	Metric Biscuits † in mm (L x W x T)	Inch Biscuits † in inches (L x W x T)	Notes
#H9	38 x 12 x 3 mm‡		Uses a smaller cutter wheel 3 mm wide.
#0	47 x 15 x 4 mm‡	1-27/32" x 5/8" x 19/128"	Standard cutter width is 4 mm or 5/32".
#10	53 x 19 x 4 mm‡	2-1/8" x 3/4" x 19/128"	

#20	56 x 23 x 4 mm‡	2-3/8" x 1" x 19/128"	One source uses 2-1/4" for length.
D	Furniture hinge	Depth of groove : 13mm	**ONLY use on biscuit joiner with Six depth setting**
S	Slide-in connector	Depth of groove : 14.7 mm	**ONLY use on biscuit joiner with Six depth setting**
S6	85 x 30 x 4 mm‡		

note: Six depth settings of Biscuit joiner (Six size biscuits with No blade change) includes #00,#10,#20,D,S,S6

† Biscuits may also be referred to as *plates* (as per the Lamello website). ‡ These data require clarification because the standard cutter width is 4 mm thus requiring the biscuit to be thinner. It is more likely that the thickness is 3.75 mm which would correspond well to the typical inch thickness (19/128" = 3.77 mm).

Note: The mm sizes were taken verbatim from the Lamello Catalogue. The inch sizes were taken verbatim from an article on plate joinery published in The Woodworker's Gazette several years ago. In general, the sizes appear to be consistent with each other given the typical tolerances used in woodworking. The usual caveats in dealing with tools and materials destined for US or European use are to be observed, of course. The most commonly used inch sizes used are #0, #10 and #20 hence their exclusive listing.

Sizes Of Porter Cable Biscuits

Size	Metric Biscuits in mm (L x W)	Inch Biscuits in inches (L x W)	Notes
#FF	30 X 13 mm	1-13/64" X 1/2"	FF = Face Frame for 1-1/2" width, and up.
#0	47 X 16 mm	1-27/32" X 5/8"	
#10	52 X 20 mm	2-3/64" X 25/32" (~3/4")	
#20	58 X 24 mm	2-9/32" (~2-1/4") X 15/16" (~1")	

Note: The sizes were taken verbatim from the Porter-Cable website.

Detail Biscuit Sizes

Detail biscuits are smaller than standard biscuits and are typically used to join smaller pieces of wood together, and offer less structural support.

Size	Metric Biscuits in mm (L x W x T)	Inch Biscuits in inches (L x W x T)	Notes
R1	16 x 5.6 x 2.4 mm	5/8" x 7/32" x 3/32"	
R2	19 x 7.1 x 2.4 mm	3/4" x 9/32" x 3/32"	
R3	25.4 x 12.7 x 2.4 mm	1" x 1/2" x 3/32"	

Blades and Depth

For most portable plate joiners, a nominal 4 inch or 100 mm diameter blade is used for the #0, #10, #20 biscuit cuts. The blade is set deeper for joining the larger biscuits. Most blades have 4, 6, or 8 teeth and fit a 7/8 inch or 22 mm arbor. The thickness of the blade is typically 0.156 to 0.160 inch or nominally 4 mm.

Bandsaw

A small portable bandsaw

Students maneuver a large laminated board through a bandsaw together

Horizontal bandsaw resawing planks at a boatyard in Hoi AN, Vietnam

Larger resaw at a Mekong delta boatyard, fitted with a 150 mm (6") blade

Bandsaw manufactured in 1911

A bandsaw uses a long sharp blade consisting of a continuous band of toothed metal rotating on opposing wheels to cut material. They are used principally in woodworking, metalworking, and lumbering, but may cut a variety of materials. Advantages include uniform cutting action as a result of an evenly distributed tooth load, and the ability to cut irregular or curved shapes like a jigsaw. The minimum radius of a curve is determined by the width of the band and its kerf. Most bandsaws have two wheels connected by a belt or chain rotating in the same plane, one of which is powered, although some may have three or four to distribute the load.

History

The idea of the band saw dates back to at least 1809, when William Newberry received a British patent for the idea, but band saws remained impractical largely because of the inability to produce accurate and durable blades using the technology of the day. Constant flexing of the blade over the wheels caused either the material or the joint welding it into a loop to fail.

Nearly 40 years passed before Frenchwoman Anne Paulin Crepin devised a welding technique overcoming this hurdle. She applied for a patent in 1846, and soon afterward sold the right to employ it to manufacturer A. Perin & Company of Paris. Combining this method with new steel alloys and advanced tempering techniques allowed Perin to create the first modern band saw blade.

The first American band saw patent was granted to Benjamin Barker of Ellsworth, Maine, in January 1836. The first factory produced and commercially available band saw in the U.S. was by a design of Paul Prybil.

Power hacksaws (with reciprocating blades) were once common in the metalworking industries, but bandsaws and cold saws have mostly displaced them.

Types

Meat Cutting

Saws for cutting meat are typically of all stainless steel construction with easy to clean features. The blades either have fine teeth with heat treated tips, or have plain or scalloped knife edges.

Metal Cutting

Bandsaws for cutting metal are available in vertical and horizontal designs. Typical band speeds range from 40 feet (12 meters) per minute to 5,000 feet (1,500 meters) per minute, however specialized bandsaws are built for friction cutting of hard metals and run band speeds to 15,000 feet per minute. Metal-cutting bandsaws are usually equipped with brushes or brushwheels to prevent chips from becoming stuck in between the blade's teeth. Systems which cool the blade are also common equipment on metal-cutting bandsaws. The coolant washes away chips and keeps the blade cool and lubricated.

19th century wood bandsaw

Horizontal bandsaws hold the workpiece stationary while the blade swings down through the cut. This configuration is used to cut long materials such as pipe or bar stock to length. Thus it is an important part of the facilities in most machine shops. The horizontal design is not useful for cutting curves or complicated shapes. Small horizontal bandsaws typically employ a gravity feed alone, retarded to an adjustable degree by a coil spring; on industrial models, the rate of descent is usually controlled by a hydraulic cylinder bleeding through an adjustable valve. When the saw is set up for a cut, the operator raises the saw, positions the material to be cut underneath the blade, and then turns on the saw. The blade slowly descends into the material, cutting it as the band blade moves. When the cut is complete, a switch is tripped and the saw automatically turns off. More sophisticated versions of this type of saw are partially or entirely automated (via PLC or CNC) for high-volume cutting of machining blanks. Such machines provide a stream of cutting fluid recirculated from a sump, in the same manner that a CNC machining center does.

A vertical bandsaw, also called a contour saw, keeps the blade's path stationary while the workpiece is moved across it. This type of saw can be used to cut out complex shapes and angles. The part may be fed into the blade manually or with a power assist mechanism. This type of metal-cutting bandsaw is often equipped with a built-in blade welder. This not only allows the operator to repair broken blades or fabricate new blades quickly, but also allows for the blade to be purposely cut, routed through the center of a part, and re-welded in order to make interior cuts. These saws are often fitted with a built-in air blower to cool the blade and to blow chips away from the cut area giving the operator a clear view of the work. This type of saw is also built in a woodworking version. The woodworking type is generally of much lighter construction and does not incorporate a power feed mechanism, coolant, or welder.

Advancements have also been made in the bandsaw blades used to cut metals. The development of new tooth geometries and tooth pitches have produced increased production rates and greater blade life. New materials and processes such as M51 steel and the cryogenic treatment of blades have produced results that were thought impossible just a few years ago. New machines have been developed to automate the welding process of bandsaw blades as well. Ideal computerized welding machines, setting and cut to length machines and contributions from other manufacturers continue to increase productivity.

Timber Cutting

Timber mills use very large bandsaws for ripping lumber; they are preferred over circular saws for ripping because they can accommodate large-diameter timber and because of their smaller kerf (cut size), resulting in less waste.

The blades are mounted on wheels with a diameter large enough not to cause metal fatigue due to flexing when the blade repeatedly changes from a circular to a straight profile. It is stretched very tight (with fatigue strength of the saw metal being the limiting factor). Bandsaws of this size need to have a deformation worked into them that counteracts the forces and heating of operation. This is called **benching**. They also need to be removed and serviced at regular intervals. Sawfilers or sawdoctors are the craftsmen responsible for this work.

The shape of the tooth gullet is highly optimized and designed by the sawyer and sawfiler. It varies according to the mill, as well as the type and condition of the wood. Frozen logs often require a frost notch ground into the gullet to break the chips. The shape of the tooth gullet is created when the blade is manufactured and its shape is automatically maintained with each sharpening. The sawfiler will need to maintain the grinding wheel's profile with periodic dressing of the wheel.

Proper tracking of the blade is crucial to accurate cutting and considerably reduces blade breakage. The first step to ensuring good tracking is to check that the two bandwheels or flywheels are co-planar. This can be done by placing a straightedge across the front of the wheels and adjusting until each wheel touches. Rotate the wheels with the blade in position and properly tensioned and check that the tracking is correct. Now install the blade guide rollers and leave a gap of about 1 mm between the back of the blade and the guide flange. The teeth of blades that have become narrow through repeated sharpening will foul the front edge of the guide rollers due to their kerf set and force the blade out of alignment. This can be remedied by cutting of a small step on the rollers' front edges to accommodate the protruding teeth. Ideally the rollers should be crowned, a configuration that assists in the proper tracking of bands and belts, at the same time allowing clearance for the set of the teeth.

Head Saws

Head saws are large bandsaws that make the initial cuts in a log. They generally have a 2 to 3 in (51 to 76 mm) tooth space on the cutting edge and *sliver teeth* on the back. Sliver teeth are non-cutting teeth designed to wipe slivers out of the way when the blade needs to back out of a cut.

Resaws

A resaw is a large bandsaw optimized for cutting timber along the grain to reduce larger sections into smaller sections or veneers. Resawing veneers requires a wide blade—commonly 2 to 3 in (51 to 76 mm)—with a small kerf to minimize waste. Resaw blades of up to 1 in (25 mm) may be fitted to a standard bandsaw.

Double Cut Saws

Double cut saws have cutting teeth on both sides. They are generally very large, similar in size to a head saw.

Construction

Feed Mechanisms

- Gravity feed saws fall under their own weight. Most such saws have a method to allow the cutting force to be adjusted, such as a movable counterbalancing weight, a coil spring with a screw-thread adjustment, or a hydraulic or pneumatic damper (speed control valve). The latter does not force the blade downwards, but rather simply limits the speed at which the saw can fall, preventing excessive feed on thin or soft parts. This is analogous to door closer hardware whose dampering action keeps the door from slamming. Gravity feed designs are common in small saws.

- Hydraulic feed saws use a positive pressure hydraulic piston to advance the saw through the work at variable pressure and rate. Common in production saws.

- Screw feed saws employ a leadscrew to move the saw.

Fall Mechanisms

- Pivot saws hinge in an arc as they advance through the work.

- Single column saws have a large diameter column that the entire saw rides up and down on, very similar to a drill press.

- Dual column saws have a pair of large columns, one on either side of the work, for very high rigidity and precision. The dual column setup is unable to make use of a miter base due to inherent design. Dual column saws are the largest variety of machine bandsaws encountered, to the point where some make use of a rotary table and X axis to perform complex cutting.

Automated Saws

Automatic bandsaws feature preset feed rate, return, fall, part feeding, and part clamping. These are used in production environments where having a machine operator per saw is not practical. One operator can feed and unload many automatic saws.

Some automatic saws rely on numerical control to not only cut faster, but to be more precise and perform more complex miter cuts.

Common Tooth Forms

- Precision blade gives accurate cuts with a smooth finish.

- Buttress blade provides faster cutting and large chip loads.

- Claw tooth blade gives additional clearance for fast cuts and soft material.

At least two teeth must be in contact with the workpiece at all times to avoid stripping off the teeth.

Chainsaw

A chainsaw (or chain saw) is a portable, mechanical saw which cuts with a set of teeth attached to a rotating chain that runs along a guide bar. It is used in activities such as tree felling, limbing, bucking, pruning, cutting firebreaks in wildland fire suppression, and harvesting of firewood. Chainsaws with specially designed bar and chain combinations have been developed as tools for use in chainsaw art and chainsaw mills. Specialist chainsaws are used for cutting concrete. Chainsaws are sometimes used for cutting ice, for example for ice sculpture and in Finland for winter swimming. Someone who uses a saw is a sawyer.

A Stihl chainsaw

Construction

A chainsaw consists of several parts:

The cutting chain seen here features the popular chipper teeth style cutting blades

Engine

Chainsaw engines are traditionally either two-stroke gasoline (petrol) internal combustion engine (usually with a cylinder volume of 30 to 120 cm^3) or an electric motor driven by a battery or electric power cord. Combustion engines today (2016) are supplied through a traditional carburetor or an electronically adjustable carburetor.

The traditional carburetor needs to be adjusted, i. e. when operating in high or low altitudes, or their fuel oil-to-gasoline ratios must be adjusted to run properly. Electrically influenced carburetors make all adjustments automatically. These systems are provided by most large chain saw producers. Husqvarna calls its "Autotune," and it is commonly standard on most saws of the 5XX saw series.

To reduce user fatigue problems, traditional carburetors can be de-vibrated (protected from vibrations) or they can be heated as well. Many saws offer a Winter and Summer mode of operation. Winter mode applies in temperatures below 0 °C / 32 °F where inside the cover a hole is opened leaving warm air to the air filter and carburetor to prevent icing. In warmer environment the hole is closed and both units are not ventilated with warm air.

To ensure clean air supply to the carburetor, chainsaw producers offer different filters with fine or less fine mesh. In clean surrounding air a less fine filter can be used, in dusty environment the other. The fine filter keeps the air clean to its optimum (i.e. 44 μm) but has the tendency to clogging. This leads the engine to die.

The engines are designed so that they may be operated in different positions, upside-down or tilted 90 degrees. Early engines died when tilting (two man saw from Dolmar, Germany from 1930 to 1937).

Drive Mechanism

Typically a centrifugal clutch and sprocket. The centrifugal clutch expands with raising spinning speed towards a drum. On this drum sits either a fixed sprocket or an exchangeable one. The clutch has three jobs to do: When the saw runs idle (typically 2500-2700 rpm) the chain does not move. When the clutch is engaged and the chain stops in the wood or another reason, it protects the engine. Most important it protects the operator in case of a kickback. Here the chain break stops the drum and the clutch releases immediately.

Clutches and drums can be in two positions: either turned outside (Husqvarna) or inside (Stihl).

Guide Bar

An elongated bar with a round end of wear-resistant alloy steel typically 40 to 90 cm (16 to 36 in) in length. An edge slot guides the cutting chain. Specialized loop-style bars, called bow bars, were also used at one time for bucking logs and clearing brush, although they are now rarely encountered due to increased hazards of operation.

All guide bars have some elements for operation:

Gauge

The lower part of the chain runs in the gauge. Here the lubrification oil is pulled by the chain to the nose. This is a very important mechanism.

Oil holes

At the end of the saw power head there are two oil holes, one on each side. These holes must match with the outlet of the oil pump. The pump presses the oil through the hole in the lower part of the gauge.

Saw bar producers provide a large variety of bars matching different saws.

Grease Holes at Bar Nose

Through this hole grease is pressed, typically each tank filling to keep the nose sprocket well lubricated.

Guide Slot

Here one or two bolts from the saw run through. The clutch cover is put on top of the bar and it is secured though this/these bolts. It depends on the size of the saw if one or two bolts are installed.

Bar Types

There are different bar types available:

- Laminated bars

These bars consist of different layers to reduce the weight of the bar.

- Solid bars

These bars are solid steel bars intended for professional use. They have commonly an exchangeable nose since the sprocket at the bar nose wears out faster at the bar.

- Safety bars

These bars are laminated bars with a small sprocket at the nose. The small nose reduces the kickback effect. Such bars are used on consumer saws.

Cutting Chain

Usually each segment in this chain (which is constructed from riveted metal sections similar to a bicycle chain, but without rollers) features small sharp cutting teeth. Each tooth takes the form of a folded tab of chromium-plated steel with a sharp angular or curved corner and two cutting edges, one on the top plate and one on the side plate. Left-handed and right-handed teeth are alternated in the chain. Chains come in varying pitch and gauge; the pitch of a chain is defined as half of the length spanned by any three consecutive rivets (e.g., 8 mm, 0.325 inch), while the gauge is the thickness of the drive link where it fits into the guide bar (e.g., 1.5 mm, 0.05 inch). Conventional "full complement" chain has one tooth for every two drive links. "Full skip" chain has one tooth for every three drive links. Built into each tooth is a depth gauge or "raker" which rides ahead of the tooth and limits the depth of cut, typically to around 0.5 mm (0.025"). Depth gauges are critical to safe chain operation. If left too high they will cause very slow cutting, if filed too low the chain will become more prone to kick back. Low depth gauges will also cause the saw to vibrate excessively. Vibration is not only uncomfortable for the operator but is also detrimental to the saw.

Tensioning Mechanism

Some way to adjust the tension in the cutting chain so that it neither binds on nor comes loose from the guide bar. The tensioner is either operated by turning a screw or a manual wheel. The tensioner is either in a lateral position underneath the exhaust or integrated in the clutch cover.

The lateral tensioner has the advantage that the clutch cover is easier to mount but the disadvantage that it is more difficult to reach nearby the bar. Tensioners through the clutch cover are easier to operate, but the clutch cover is more difficult to attach.

When turning the screw, a hook in a bar hole moves the bar either out (tensioning) or in, making the chain loose. Tension is right when it can be moved easily by hand and not hanging loose from the bar. When tensioning, hold the bar nose up and pull the bar nuts tight. Otherwise the chain might derail.

The underside of each link features a small metal finger called a "drive link" (also DL) which locates the chain on the bar, helps to carry lubricating oil around the bar, and engages with the engine's drive sprocket inside the body of the saw. The engine drives the chain around the track by a centrifugal clutch, engaging the chain as engine speed increases under power, but allowing it to stop as the engine speed slows to idle speed.

Dramatic improvements, chainsaw safety devices and overall design have taken place since the chainsaw's invention, saving many lives and preventing countless serious injuries. These include chainbrake systems, better chain design and anti-vibration systems.

As chainsaw carving has become more popular, chainsaw manufacturers are making special short, narrow-tipped bars for carving. These are called "quarter tipped," "nickel tipped" or "dime tipped" bars, based on the size of the round tip. Chainsaw manufacturer Echo sponsors a carving series, as well as carvers such as former Runaways singer Cherie Currie. Some chainsaws such as the RedMax G3200 CV are built specifically for carving applications.

Safety Features

Today's chainsaws show all a number of safety features to protect the operator. All these features are not a 100% guarantee that the operator is not harmed. The best protection, even still, is experience.

- Chain break

The chain break is located in the clutch cover. Here a band tensions around the Clutch drum stopping the chain within milliseconds. The chain break is released by the upper handle with the hand or wrist. The break is intended to be used in kick-back moments.

- Chain catcher

The chain catcher is located between the saw body and the clutch cover. In most cases it looks as a hook made in aluminum. It is used to stop the chain when it derails from the bar and shortens the length of the chain. When derailing the chain swings from underneath the saw towards the operator. The shorting prevents hitting the operator, but it hits the rear handle gurard.

- Rear handle guard

The rear handle guard protects the hand of the operator when the chain derails.

- Chain

Some chains show safety features as safety links as on micro chisel saws. These links keep saw close the gap between two cutting links and lift the chain when the space at the safety link is full with saw chips. This lifts the chain and let it cut slower.

Maintenance

Two-stroke chainsaws require about 2–5% of oil in the fuel to lubricate the motor, while the motor in electrical chain-saws is normally lubricated for life. Most modern gas operated saws today require a fuel mix of 2% (1:50). Regular gas from most gas stations contain 5 to 10% ethanol which can result in problems of the equipment. Ethanol dissolves plastic, rubber and other material. This leads to problems especially on older equipment. A workaround of this problem is to run fresh fuel only and run the saw dry at the end of the work.

Logging near Apiary, Oregon

Separate *chain oil* or *bar oil* is used for the lubrication of the bar and chain on all types of chain-saw. The chain oil is depleted quickly because it tends to be thrown off by chain centrifugal force, and it is soaked up by sawdust. On two-stroke chainsaws the chain oil reservoir is usually filled up at the same time as refuelling. The reservoir is normally large enough to provide sufficient chain oil between refuelling. Lack of chain-oil, or using an oil of incorrect viscosity, is a common source of damage to chain-saws, and tends to lead to rapid wear of the bar, or the chain seizing or coming off the bar. In addition to being quite thick, chain oil is particularly sticky (due to "tackifier" additives) to reduce the amount thrown off the chain. Although motor oil is a common emergency substitute, it is lost even faster and so leaves the chain under-lubricated.

Chain oil is either non-biodegradeable or degradable. Professionals have to use biode-gradeable oil in Germany by law.

The oil is pumped from a small pump to a hole in the bar. From here the lower ends of each chain drive link take a portion of the oil into the gauge towards the bar nose. Pump outlet and bar hole must be aligned. Since the bar is moving out and inwards depending on the chain length, the oil outlet on the saw side has a banana style long shape.

Chains must be kept sharp to perform well. They become blunt rapidly if they touch soil, metal or stones. When blunt, they tend to produce powdery sawdust, rather than the longer, clean shavings characteristic of a sharp chain; a sharp saw also needs very little force from the operator to push it into the cut. Special hardened chains (made with tungsten carbide) are used for applications where soil is likely to contaminate the cut, such as for cutting through roots.

A clear sign of a blunt chain are vibrations of the saw. A sharp chain pulls itself into the wood without pressing on the saw.

The air intake filter tends to clog up with sawdust. This must be cleaned from time to time, but is not a problem during normal operation.

Safety

A chainsaw operator wearing full safety gear using a gasoline-powered chain saw

Despite safety features and protective clothing, injuries can still arise from chainsaw use, from the large forces involved in the work, from the fast-moving, sharp chain, or from the vibration and noise of the machinery.

A common accident arises from *kickback*, when a chain tooth at the tip of the guide bar catches on wood without cutting through it. This throws the bar (with its moving chain) in an upward arc toward the operator which can cause serious injury or even death.

Another dangerous situation occurs when heavy timber begins to fall or shift before a cut is complete — the chainsaw operator may be trapped or crushed. Similarly, timber falling in an unplanned direction may harm the operator or other workers, or an operator working at a height may fall or be injured by falling timber.

Like other hand-held machinery, the operation of chainsaws can cause vibration white finger, tinnitus or industrial deafness. Theses symptoms were very common when such equipment was not de-vibrated. On today's equipment there are damping elements (in rubber or steel spring) lowering these risks. Heated handles are an additional help.

The risks associated with chainsaw use mean that protective clothing such as chainsaw boots, chainsaw trousers and hearing protectors are normally worn while operating them, and many jurisdictions require that operators be certified or licensed to work with chainsaws. Injury can also result if the chain breaks during operation due to poor maintenance or attempting to cut inappropriate materials.

Gasoline-powered chainsaws expose operators to harmful carbon monoxide (CO) gas, especially indoors or in partially enclosed outdoor areas.

Drop starting, or turning on a chainsaw by dropping it with one hand while pulling the starting cord with the other, is a safety violation in most states in the U.S. Keeping both hands on the saw for stability is essential for safe chainsaw use.

Safe and effective chainsaw and crosscut use on Federally-administered public lands within the United States has been codified since July 19th, 2016 in the publication of the *Final Directive for National Saw Program* issued by the United States Forest Service, USDA which specifies the training, testing, and certification process for employees as well as for unpaid volunteers who operate chainsaws within public lands.

The new directive specifies Forest Service Manual (FSM) 2358 (PDF) which covers classification of sawyers, their Personal Protective Equipment (PPE) and numerous other aspects of required safety training and behavior when operating chainsaws or crosscut saws on Federally-administered public lands.

Working Techniques

Chainsaw training is designed to provide working technical knowledge and skills to safely operate the equipment.

- Sizeup – This is scouting and planning safe cuts for the felling direction, danger zones, and retreat paths, before starting the saw. The tree's location relative to other objects, support, and tension determines a safe fall, splits off or if the saw will jam. Several factors to consider are: tree lean and bend, wind direction, branch arrangement, snow load, obstacles and damaged, rotting tree parts, which might behave unexpectedly when cut. A tree may have to fall in its natural direction if it's too dangerous or impossible to fell in a desired direction. The aim is for the tree to fall safely for limbing and cross-cutting the log. The goal is to avoid having the tree fall on another tree or obstacle.

- Felling – After clearing the tree's base undergrowth for the retreat path and the felling direction; felling is properly done with three main cuts. To control the fall, the directional cut line should run 1/4 of the tree diameter to make a 45-degree wedge, which should be 90 degrees to the felling direction and perfectly horizontal. Make the top cut first then, the bottom cut is made to form the directional cut line at the wedge point. A narrow or nonexistent hinge lessens felling direction control. From the opposite side of the wedge, plan to finish the final felling cut 1/10 of the tree diameter from the direction cut line. The felling cut is made horizontally and slightly (1.5–2 inches; 5.1 cm) above the bottom cut. When the hinge is properly set, the felling cut will begin the fall in the desired direction. A sitback is when a tree moves back opposite the intended direction. Placing a wedge in the felling cut can prevent a sitback from pinching the saw.

- Freeing – Working a badly fallen tree that may have become trapped in other trees. Working out maximum tension locations to decide the safest way to release tension, and a winch may be needed in complicated situations. To avoid cutting straight through a tree in tension, one or two cuts at the tension point of sufficient depth to reduce tension may be necessary. After tension releases, cuts are made outside the bend.

- Limbing – Cutting the branches off the log. The operator must be able to properly reach the cut to avoid kickback.

- Bucking – Cross-cutting the felled log into sections. Setup is made to avoid binding the chainsaw within the changing log tensions and compressions. Safe bucking is started at the log highside and then sections worked offside, toward the butt end. The offside log falls and allows for gravity to help prevent binds. Watching the log's kerf movement while cutting, helps to indicate binds. Additional equipment (lifts, bars, wedges and winches) and special cutting techniques can help prevent binds.

- Binds – This is when the chainsaw is at risk or is stuck in the log compression. A log bound chainsaw is not safe, and must be carefully removed to prevent equipment damage.

- o   Top bind – The tension area on log bottom, compression on top.

- o   Bottom bind – The tension area on log top, compression on bottom.

- o   Side bind – Sideways pressure exerted on log.

- o   End bind – Weight compresses the log's entire cross section.

- Brushing and Slashing – This is quickly clearing small trees and branches under 5 inches diameter. A hand piler may follow along to move out debris.

History

The origin is debated, but a chainsaw-like tool was made around 1830 by the German orthopaedist Bernhard Heine. This instrument, the osteotome, had links of a chain carrying small cutting teeth with the edges set at an angle; the chain was moved around a guiding blade by turning the handle of a sprocket wheel. As the name implies, this was used to cut bone. The prototype of the chain saw familiar today in the timber industry was pioneered in the late 18th century by two Scottish doctors, John Aitken and James Jeffray, for symphysiotomy and excision of diseased bone respectively. The chain hand saw, a fine serrated link chain which cut on the concave side, was invented around 1783-1785. It was illustrated in Aitken's Principles of Midwifery or Puerperal Medicine (1785) and used by him in his dissecting room. Jeffray claimed to have conceived the idea of the chain saw independently about that time but it was 1790 before he was able to have it produced. In 1806, Jeffray published Cases of the Excision of Carious Joints by H. Park and P. F. Moreau with Observations by James Jeffray M.D.. In this communication he translated Moreau's paper of 1803. Park and Moreau described successful excision of diseased joints, particularly the knee and elbow. Jeffray explained that the chain saw would allow a smaller wound and protect the adjacent neurovascular bundle. While a heroic concept, symphysiotomy had too many complications for most obstetricians but Jeffray's ideas became accepted, especially after the development of anaesthetics. Mechanised versions of the chain saw were developed but in the later 19th Century, it was superseded in surgery by the Gigli twisted wire saw. For much of the 19th century, however, the chain saw was a useful surgical instrument.

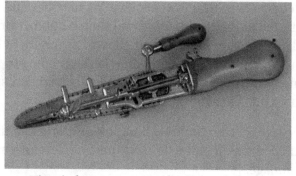

Historical osteotome, a medical bone chainsaw

Typical of the earliest chainsaws, this Dolmar saw is operated by two men.

The first portable chainsaw was developed and patented in 1918 by Canadian mill-wright James Shand. After he allowed his rights to lapse in 1930 his invention was further developed by what became the German company Festo in 1933. The company now operates as Festool producing portable power tools. Other important contributors to the modern chainsaw are Joseph Buford Cox and Andreas Stihl; the latter patented and developed an electrical chainsaw for use on bucking sites in 1926 and a gasoline-powered chainsaw in 1929, and founded a company to mass-produce them. In 1927, Emil Lerp, the founder of Dolmar, developed the world's first gasoline-powered chainsaw and mass-produced them.

McCulloch electric chainsaw

World War II interrupted the supply of German chain saws to North America, so new manufacturers sprang up including Industrial Engineering Ltd (IEL) in 1947, the fore-runner of Pioneer Saws. Ltd and part of Outboard Marine Corporation, the oldest manufacturer of chainsaws in North America.

McCulloch in North America started to produce chainsaws in 1948. The early models were heavy, two-person devices with long bars. Often chainsaws were so heavy that they had wheels like dragsaws. Other outfits used driven lines from a wheeled power unit to drive the cutting bar.

After World War II, improvements in aluminum and engine design lightened chain-saws to the point where one person could carry them. In some areas the skidder (chain-saw) crews have been replaced by the feller buncher and harvester.

Chainsaws have almost entirely replaced simple man-powered saws in forestry. They come in many sizes, from small electric saws intended for home and garden use, to large "lumberjack" saws. Members of military engineer units are trained to use chainsaws as are firefighters to fight forest fires and to ventilate structure fires.

Cutting Stone, Concrete and Brick

Special chainsaws can cut concrete, brick and natural stone. These use similar chains to ordinary chainsaws, but with cutting edges embedded with diamond grit. They may use gasoline or hydraulic power, and the chain is lubricated with water, because of high friction and to remove stone-dust. The machine is used in construction, for example in cutting deep square holes in walls or floors, in stone sculpture for removing large chunks of stone during pre-carving, by fire departments for gaining access to buildings and in restoration of buildings and monuments, for removing parts with minimal damage to the surrounding structure. More recently concrete chainsaws with electric motors of 230 volts have also been developed.

A chainsaw cutting concrete. The hose supplies cooling water.

Because the material to be cut is non-fibrous, there is much less chance of kickback. Therefore, the most-used method of cutting is plunge-cutting, by pushing the tip of the blade into the material. With this method square cuts as small as the blade width can be achieved. Pushback can occur if a block shifts when nearly cut through and pinches the blade, but overall the machine is less dangerous than a wood-cutting chainsaw.

Hand Tool

A hand tool is any tool that is not a power tool – that is, one powered by hand (manual labour) rather than by an engine. Some examples of hand tools are garden forks, secateurs, rakes, hammers, spanners, pliers, screwdrivers and chisels. Hand tools are generally less dangerous than power tools.

Brief History

Hand tools have been used by humans since the stone age when stones were used for hammering and cutting. During the bronze age tools were made by casting the copper and tin alloys. Bronze tools were sharper and harder than those made of stone. During the iron age iron replaced bronze, and tools became even stronger and more durable. The Romans developed tools during this period which are similar to those being produced today. In the period since the industrial revolution, the manufacture of tools has transitioned from being craftsman made to being factory produced.

A large collection of British hand tools dating from 1700 to 1950 is held by St Albans Museums. Most of the tools were collected by Raphael Salaman (1906–1993) who wrote two classic works on the subject: *Dictionary of Woodworking Tools* and *Dictionary of Leather-working Tools*.

General Tool Categories

The American Industrial Hygiene Association gives the following categories of hand tools: wrenches, pliers, cutters, striking tools, struck or hammered tools, screwdrivers, vises, clamps, snips, saws, drills and knives.

Hacksaw

A hacksaw is a fine-toothed saw, originally and principally made for cutting metal. They can also cut various other materials, such as plastic and wood; for example, plumbers and electricians often cut plastic pipe and plastic conduit with them. There are hand saw versions and powered versions (power hacksaws). Most hacksaws are hand saws with a C-shaped frame that holds a blade under tension. Such hacksaws have a handle, usually a pistol grip, with pins for attaching a narrow disposable blade. The frames may also be adjustable to accommodate blades of different sizes. A screw or other mechanism is used to put the thin blade under tension. Panel hacksaws forgo the frame and instead have a sheet metal body; they can cut into a sheet metal panel further than a frame would allow. These saws are no longer commonly available, but hacksaw blade holders enable standard hacksaw blades to be used similarly to a keyhole saw or pad saw. Power tools including nibblers, jigsaws, and angle grinders fitted with metal-cutting blades and discs are now used for longer cuts in sheet metals.

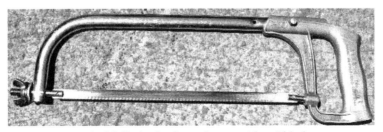

Typical full-size hacksaw frame, with 12" blade

On hacksaws, as with most frame saws, the blade can be mounted with the teeth facing toward or away from the handle, resulting in cutting action on either the push or pull stroke. In normal use, cutting vertically downwards with work held in a bench vice, hacksaw blades should be set to be facing forwards. Some frame saws, including Fret Saws and Piercing Saws, have their blades set to be facing the handle because they are used to cut by being pulled down against a horizontal surface.

Design

History

While saws for cutting metal had been in use for many years, significant improvements in longevity and efficiency were made in the 1880s by George N. Clemson, a founder of Clemson Bros., Inc of Middletown, New York, United States,. Clemson conducted tests which involved changing the dimensions, shapes of teeth, styles of set, and variable heat treatments of blades. Clemson claimed enormous improvements to the cutting ability of blades and built a major industrial operation manufacturing hacksaw blades sold under the trade name Star Hack Saw. In 1898, Clemson was granted US Patent 601947, which details various improvements in the hacksaw.

Blades

Blades are available in standardized lengths, 10 or 12 inches (254 or 305 mm) for a standard hand hacksaw. "Junior" hacksaws are 6 inches (152 mm) long. Powered hacksaws may use large blades in a range of sizes, or small machines may use the same hand blades.

Junior hacksaw

The pitch of the teeth can be anywhere from fourteen to thirty-two teeth per inch (tpi) for a hand blade, with as few as three tpi for a large power hacksaw blade. The blade chosen is based on the thickness of the material being cut, with a minimum of three teeth in the material. As hacksaw teeth are so small, they are set in a "wave" set. As for other saws they are set from side to side to provide a kerf or clearance when sawing, but the set of a hacksaw changes gradually from tooth to tooth in a smooth curve, rather than alternate teeth set left and right.

Hacksaw blades are normally quite brittle, so care needs to be taken to prevent brittle fracture of the blade. Early blades were of carbon steel, now termed 'low alloy' blades, and were relatively soft and flexible. They avoided breakage, but also wore out rapidly. Except where cost is a particular concern, this type is now obsolete. 'Low alloy' blades are still the only type available for the Junior hacksaw, which limits the usefulness of this otherwise popular saw.

For several decades now, hacksaw blades have used high speed steel for their teeth, giving greatly improved cutting and tooth life. These blades were first available in the 'All-hard' form which cut accurately but were extremely brittle. This limited their practical use to benchwork on a workpiece that was firmly clamped in a vice. A softer form of high speed steel blade was also available, which wore well and resisted breakage, but was less stiff and so less accurate for precise sawing. Since the 1980s, bi-metal blades have been used to give the advantages of both forms, without risk of breakage. A strip of high speed steel along the tooth edge is electron beam welded to a softer spine. As the price of these has dropped to be comparable with the older blades, their use is now almost universal.

Hacksaw blade specifications: The most common blade is the 12 inch or 300 mm length. Hacksaw blades have two holes near the ends for mounting them in the saw frame and the 12 inch / 300 mm dimension refers to the center to center distance between these mounting holes.

12 Inch Blade

   Hole to hole: 11 7/8 inches / 300. mm

   Overall blade length: 12 3/8 inches / 315 mm (not tightly controlled)

   Mounting Hole diameter: 9/64 to 5/32 inch / 3.6 to 4 mm (not tightly controlled)

   Blade Width: 7/16 to 33/64 inch / 11 to 13 mm (not tightly controlled)

   Blade Thickness: 0.020 to 0.027 inches / 0.5 to 0.70 mm (varies with tooth pitch and other factors)

The kerf produced by the blades is somewhat wider than the blade thickness due to the set of the teeth. It commonly varies between 0.030 and 0.063 inches / 0.75 and 1.6 mm depending on the pitch and set of the teeth.

The 10 inch blade is also fairly common and all the above dimensions apply except for the following:

   Hole to Hole: 9 7/8 inches / 250 mm

   Overall blade length: 10 3/8 inches / 265 mm (not tightly controlled)

Variants

A *panel hacksaw* has a frame made of a deep, thin sheet aligned behind the blade's kerf, so that the saw could cut into panels of sheet metal without the length of cut being restricted by the frame. The frame follows the blade down the kerf into the panel.

A panel hacksaw

Junior hacksaws are a small version with a half-size blade. Like coping saws, the blade has pins that are held by notches in the frame. Although potentially a useful tool for a toolbox or in confined spaces, the quality of blades in the Junior size is restricted and they are only made in the simple low alloy steels, not HSS. This restricts their usefulness.

An electric hacksaw

A *power hacksaw* (or *electric hacksaw*) is a type of hacksaw that is powered either by its own electric motor or connected to a stationary engine. Most power hacksaws are stationary machines but some portable models do exist; the latter (with frames) have been displaced to some extent by reciprocating saws such as the Sawzall, which accept blades with hacksaw teeth. Stationary models usually have a mechanism to lift up the saw blade on the return stroke and some have a coolant pump to prevent the saw blade from overheating.

Power hacksaws are not as commonly used in the metalworking industries as they once were. Bandsaws and cold saws have mostly displaced them. While stationary electric hacksaws are not very common, they are still produced. Power hacksaws of the type powered by stationary engines and line shafts, like other line-shaft-powered machines, are now rare; museums and antique-tool hobbyists still preserve a few of them.

Card Scraper

A card scraper is a woodworking shaping and finishing tool. It is used to manually remove small amounts of material and excels in tricky grain areas where hand planes would cause tear out. Card scrapers are most suitable for working with hardwoods, and can be used instead of sandpaper. Scraping produces a cleaner surface than sanding; it does not clog the pores of the wood with dust, and does not leave a fuzz of torn fibers, as even the finest abrasives will do.

Card scraper in use

Types of Card Scrapers

Card scrapers are available in a range of shapes and sizes, the most common being a rectangular shape approximately 3"x 6". Another common configuration is the *goose-neck* scraper, which has a shape resembling a french curve and is useful for scraping curved surfaces. For scraping convex shapes such as violin fingerboards, small flexible rectangular scrapers are useful.

Various scrapers, with burnisher

Scrapers are normally made from high carbon steel. There are many manufacturers who provide scrapers in a wide variety of styles. Many woodworkers prefer to make their own card scrapers by cutting them from old hand saw blades.

Card scrapers are sometimes used in working with ceramics, where they may substitute for the more traditional wooden rib. The scraper is also useful for trimming damp or dry clay. Such a scraper, the nearest one to the camera, is shown in the image at right.

Turning The Burr

The cutting component of a card scraper is the burred edge of the scraper. The burr is a sharp hook of metal which is turned on the edge of the scraper by burnishing with a steel rod. A file or sharpening stone is used to joint the edge of the scraper before burnishing. Cabinet makers typically joint the edge square, or at a right angle to the face of the scraper, which allows a fine burr to be turned on both sides. Luthiers often use a beveled edge, which allows a more aggressive burr, but on one side only.

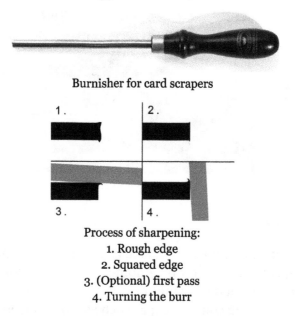

Burnisher for card scrapers

Process of sharpening:
1. Rough edge
2. Squared edge
3. (Optional) first pass
4. Turning the burr

There are a number of variations in the process of turning the burr, the choice of which is down to personal preference. However the basic concept is that the burnishing rod is held against the edge of the scraper at a slight angle and drawn along the edge a number of times until the burr is created. One variation involves holding the rod flat against the side of the scraper for the first few passes to create a burr pointing away from the edge and this is then rolled back down by drawing the rod along perpendicular to the scraper.

With the burr properly turned, a scraper will produce miniature shavings resembling those from a plane. If only dust is produced, either the scraper is not sharp, or not being used correctly.

Use

Leveling cello arching with card scraper

In use, the card scraper is grasped by the fingers, one hand on either side, and tilted in the direction of travel. The degree of tilt controls the effective angle of the cutting edge, which is the burr. Practice is required to find the correct angle. In addition to the tilt, the user will usually flex the scraper slightly by holding it at the edges with the fingers and pressing in the centre with the thumbs. The slight flex causes the corners of the scraper to lift slightly thus preventing them from marring the workpiece.

The scraper may be drawn towards the user or pushed away, although some woodworkers claim that pulling the scraper towards one can cause it to dip, resulting in an uneven surface.

The main drawback of a card scraper is that the usage is a difficult skill to master. It takes a bit of practice to turn the burr correctly and then the scraper must be presented to the work in the correct fashion. Also, due to the friction created in use, the card scraper can become hot enough to burn fingers.

Some woodworkers use the term "cabinet scraper" instead of "card scraper" for this tool. A separate tool, also called a cabinet scraper, differs in having a beveled edge like a plane.

A variation of the card scraper is the scraper plane which consists of a hand plane body in which a card scraper is mounted. There are also various styles of holder which can be used for this purpose. These simplify the use of the card scraper and alleviate the problems associated with heat.

Burnishing jigs and other accessories are available to assist in turning the burr, which makes the card scraper more accessible to the novice.

Measuring Rod

Gudea of Lagash with measuring rod and surveyors tools

Graeco-Egyptian God Serapis with measuring rod

A measuring rod is a tool used to physically measure lengths and survey areas of various sizes. Most measuring rods are round or square sectioned, however they can be flat boards. Some have markings at regular intervals. It is likely that the measuring rod was used before the line, chain or steel tapes used in modern measurement.

History

Ancient Sumer

The oldest preserved measuring rod is a copper-alloy bar which was found by the German Assyriologist Eckhard Unger while excavating at Nippur (pictured below). The bar dates from c. 2650 BC. and Unger claimed it was used as a measurement standard. This irregularly formed and irregularly marked *graduated rule* supposedly defined the *Sumerian cubit* as about 518.5 mm or 20.4 inches, although this does not agree with other evidence from the statues of Gudea from the same region, five centuries later.

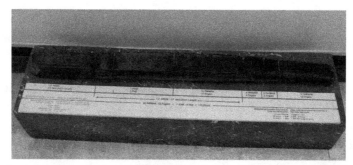

Nippur cubit, graduated specimen of an ancient measure from Nippur, Mesopotamia (3rd millennium B.C.) – displayed in the Archeological Museum of Istanbul (Turkey).

Ancient East Asia

Measuring rods for different purposes and sizes (construction, tailoring and land survey) have been found from China and elsewhere dating to the early 2nd millennium B.C.E.

Ancient Egypt

Cubit-rods of wood or stone were used in Ancient Egypt. Fourteen of these were described and compared by Lepsius in 1865. Flinders Petrie reported on a rod that shows a length of 520.5 mm, a few millimetres less than the Egyptian cubit. A slate measuring rod was also found, divided into fractions of a Royal Cubit and dating to the time of Akhenaten.

Further cubit rods have been found in the tombs of officials. Two examples are known from the tomb of Maya—the treasurer of the 18th dynasty pharaoh Tutankhamun—in Saqqara. Another was found in the tomb of Kha (TT8) in Thebes. These cubits are ca 52.5 cm long and are divided into seven palms, each palm is divided into four fingers and the fingers are further subdivided. Another wooden cubit rod was found in Theban tomb TT40 (Huy) bearing the throne name of Tutankhamun (Nebkheperure).

Cubit rod from the Turin Museum.

Egyptian measuring rods also had marks for the Remen measurement of approximately 370mm, used in construction of the Pyramids.

Ancient Europe

An oak rod from the Iron Age fortified settlement at Borre Fen in Denmark measured 53.15 inches (135.0 cm), with marks dividing it up into eight parts of 6.64 inches (16.9 cm), corresponding quite closely to half a Doric Pous. A hazel measuring rod recovered from a Bronze Age burial mound in Borum Eshøj, East Jutland by P. V. Glob in 1875 measured 30.9 inches (78 cm) corresponding remarkably well to the traditional Danish foot. The megalithic structures of Great Britain has been hypothesized to have been built by a "Megalithic Yard", though some authorities believe these structures have been measured out by pacing. Several tentative Bronze age bone fragments have been suggested as being parts of a measuring rod for this hypothetical measurement.

Roman Empire

Large public works and imperial expansion, particularly the large network of Roman roads and the many milecastles, made the measuring rod an indispensable part of both the military and civilian aspects of Roman life. Republican Rome used several measures, including the various Greek feet measurements and the Oscan foot of 23,7 cm. Standardisation was introduced by Agrippa in 29 BC, replacing all previous measurements by a Roman foot of 29.6 cm, which became the foot of Imperial Rome.

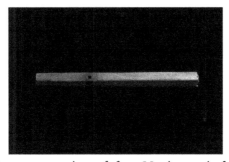

Fragment of a Roman measuring rod, from Musée romain de Lausanne-Vidy

The Roman measuring rod was 10 Roman feet long, and hence called a *decempeda*, Latin for 'ten feet'. It was usually of square section capped at both ends by a metal shoe, and painted in alternating colours. Together with the *groma* and *Dioptra* the *decempeda* formed the basic kit for the Roman surveyors. The measuring rod is frequently found depicted in Roman art showing the surveyors at work. A shorter folding yardstick one Roman foot long is known from excavations of a Roman fort in Niederburg, Germany.

Middle Ages

In the Middle Ages, bars were used as standards of length when surveying land. These bars often used a unit of measure called a rod, of length equal to 5.5 yards, 5.0292 metres, 16.5 feet, or $\frac{1}{320}$ of a statute mile. A rod is the same length as a *perch* or a *pole*. In Old English, the term *lug* is also used. The length is equal to the standardized length of the ox goad used for teams of eight oxen by medieval English ploughmen. The lengths of the perch (one rod unit) and *chain* (four rods) were standardized in 1607 by Edmund Gunter. The rod unit was still in use as a common unit of measurement in the mid-19th century, when Henry David Thoreau used it frequently when describing distances in his work *Walden*.

In Culture

Iconography

Two statues of Gudea of Lagash in the Louvre depict him sitting with a tablet on his lap, upon which are placed surveyors tools including a measuring rod.

The sun-god Shamash holding a ring of coiled rope and a rod

Measuring rod and coiled rope depicted in the Code of Hammarubi

Seal 154 recovered from Alalakh, now in the Biblioteque Nationale show a seated figure with a wedge shaped measuring rod.

The Tablet of Shamash recovered from the ancient Babylonian city of Sippar and dated to the 9th century BC shows Shamash, the Sun God awarding the measuring rod and coiled rope to newly trained surveyors.

A similar scene with measuring rod and coiled rope is shown on the top part of the diorite stele above the Code of Hammurabi in the Louvre, Paris, dating to ca. 1700 BC.

The "measuring rod" or tally stick is common in the iconography of Greek Goddess Nemesis.

The Graeco-Egyptian God Serapis is also depicted in images and on coins with a measuring rod in hand and a vessel on his head.

The most elaborate depiction is found on the Ur-Nammu-stela, where the winding of the cords has been detailed by the sculptor. This has also been described as a "staff and a chaplet of beads".

Mythology

The myth of Inanna's descent to the nether world describes how the goddess dresses and prepares herself:

"She held the lapis-lazuli measuring rod and measuring line in her hand."

Lachesis in Greek mythology was one of the three Moirai (or Fates) and "allotter" (or drawer of lots). She measured the thread of life allotted to each person with her measuring rod. Her Roman equivalent was *Decima* (the 'Tenth').

Varuna in the Rigveda, is described as using the Sun as a measuring rod to lay out space in a creation myth. W. R. Lethaby has commented on how the measurers were seen as solar deities and noted how Vishnu *"measured the regions of the Earth"*.

Bible

Measuring rods or reeds are mentioned many times in the Bible.

A measuring rod and line are seen in a vision of Yahweh in Ezekiel 40:2-3:

In visions of God he took me to the land of Israel and set me on a very high mountain, on whose south side were some buildings that looked like a city. He took me there, and I saw a man whose appearance was like bronze; he was standing in the gateway with a linen cord and a measuring rod in his hand.

Another example is Revelation 11:1:

I was given a reed like a measuring rod and was told, "Go and measure the temple of God and the altar, and count the worshipers there".

The measuring rod also appears in connection with foundation stone rites in Revelation 20:14-15:

And the wall of the city had twelve foundation stones, and on them were the twelve names of the twelve apostles of the Lamb. The one who spoke with me had a gold measuring rod to measure the city, and its gates and its wall.

Hammer

A hammer is a tool or device that delivers a blow (a sudden impact) to an object. Most hammers are hand tools used to drive nails, fit parts, forge metal, and break apart objects. Hammers vary in shape, size, and structure, depending on their purposes.

Hammers are basic tools in many trades. The usual features are a head (most often made of steel) and a handle (also called a helve or haft). Most hammers are hand tools, but there are also many powered versions, called power hammer (such as steam hammers and trip hammers) for heavier uses, such as forging.

Some hammers have other names, such as *sledgehammer*, *mallet* and *gavel*. The term "hammer" also applies to other devices that deliver blows, such as the hammer of a firearm or the hammer of a piano.

History

The use of simple hammers dates to about 2,600,000 BCE when various shaped stones were used to strike wood, bone, or other stones to break them apart and shape them. Stones attached to sticks with strips of leather or animal sinew were being used as hammers with handles by about 30,000 BCE during the middle of the Paleolithic Stone Age.

The hammer's archeological record shows that it may be the oldest tool for which definite evidence exists of its early existence.

A stone hammer found in Dover Township, Minnesota dated to 8000–3000 BCE, the North American Archaic period

Stone tapping hammer

Perforated hammer head of stone

Ancient Greek bronze sacrificial hammer, 7th century BCE, from Dodona

16th-century claw hammer; detail from Dürer's *Melencolia I* (c. 1514)

Construction and Materials

A traditional hand-held hammer consists of a separate head and a handle, fastened together by means of a special wedge made for the purpose, or by glue, or both. This two-piece design is often used, to combine a dense metallic striking head with a non-metallic mechanical-shock-absorbing handle (to reduce user fatigue from repeated strikes). If wood is used for the handle, it is often hickory or ash, which are tough and long-lasting materials that can dissipate shock waves from the hammer head. Rigid fiberglass resin may be used for the handle; this material does not absorb water or decay, but does not dissipate shock as well as wood.

A loose hammer head is hazardous because it can literally "fly off the handle" when in use, becoming a dangerous uncontrolled missile. Wooden handles can often be replaced when worn or damaged; specialized kits are available covering a range of handle sizes and designs, plus special wedges for attachment.

Some hammers are one-piece designs made primarily of a single material. A one-piece metallic hammer may optionally have its handle coated or wrapped in a resilient material such as rubber, for improved grip and reduced user fatigue.

The hammer head may be surfaced with a variety of materials, including brass, bronze, wood, plastic, rubber, or leather. Some hammers have interchangeable striking surfaces, which can be selected as needed or replaced when worn out.

Designs and Variations

A large hammer-like tool is a *maul* (sometimes called a "beetle"), a wood- or rubber-headed hammer is a *mallet*, and a hammer-like tool with a cutting blade is usually called a *hatchet*. The essential part of a hammer is the head, a compact solid mass that is able to deliver a blow to the intended target without itself deforming. The impacting surface of the tool is usually flat or slightly rounded; the opposite end of the impacting mass may have a ball shape, as in the ball-peen hammer. Some upholstery hammers have a magnetized face, to pick up tacks. In the hatchet, the flat hammer head may be secondary to the cutting edge of the tool.

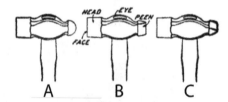

The parts of a hammer are the *face*, *head* (includes the *bell* and *neck*, which are not labelled), *eye* (where the *handle* fits into), *peen* (also spelled pein and pane). The side of a hammer is the *cheek* and some hammers have *straps* that extend down the handle for strength. Shown here are: A. Ball-peen hammer B. Straight-peen hammer C. Cross-peen hammer

The claw of a carpenter's hammer is frequently used to remove nails.

The impact between steel hammer heads and the objects being hit can create sparks, which may ignite flammable or explosive gases. These are a hazard in some industries such as underground coal mining (due to the presence of methane gas), or in other hazardous environments such as petroleum refineries and chemical plants. In these environments, a variety of non-sparking metal tools are used, primarily made of aluminium or beryllium copper. In recent years, the handles have been made of durable plastic or rubber, though wood is still widely used because of its shock-absorbing qualities and repairability.

Hand-powered Hammers

- Ball-peen hammer, or mechanic's hammer

- Boiler scaling hammer

- Brass hammer, also known as non-sparking hammer or spark-proof hammer and used mainly in flammable areas like oil fields

- Carpenter's hammer (used for nailing), such as the framing hammer and the claw hammer, and pinhammers (ball-peen and cross-peen types)

- Cow hammer – sometimes used for livestock slaughter, a practice now deprecated due to animal welfare objections

- Cross-peen hammer, having one round face and one wedge-peen face.

- Dead blow hammer delivers impact with very little recoil, often due to a hollow head filled with sand, lead shot or pellets

- Drilling hammer – a short handled sledgehammer originally used for drilling in rock with a chisel. The name usually refers to a hammer with a 2-to-4-pound (0.91 to 1.81 kg) head and a 10-inch (250 mm) handle, also called a "single-jack" hammer because it was used by one person drilling, holding the chisel in one hand and the hammer in the other. In modern usage, the term is mostly interchangeable with "engineer's hammer", although it can indicate a version with a slightly shorter handle.

- Engineer's hammer, a short-handled hammer, originally an essential components of a railroad engineer's toolkit for working on steam locomotives. Typical weight is 2–4 lbs (0.9–1.8 kg) with a 12–14 inch (30–35 cm) handle. Originally these were often cross-peen hammers, with one round face and one wedge-peen face, but in modern usage the term primarily refers to hammers with two round faces.

- Gavel, used by judges and presiding authorities to draw attention

- Geologist's hammer or rock pick

- Joiner's hammer, or Warrington hammer

- Knife-edged hammer, its properties developed to aid a hammerer in the act of slicing whilst bludgeoning

- Lathe hammer (also known as a lath hammer, lathing hammer, or lathing hatchet), a tool used for cutting and nailing wood lath, which has a small hatchet blade on one side (with a small, lateral nick for pulling nails) and a hammer head on the other

- Lump hammer, or club hammer

- Mallets, including versions made with hard rubber or rolled sheets of rawhide

- Railway track keying hammer

- Rock climbing hammer

- Rounding hammer Blacksmith or farrier hammer. Round face generally for moving or drawing metal and flat for "planishing" or smoothing out the surface marks.

- Sledge hammer

- Soft-faced hammer

- Splitting maul

- Stonemason's hammer

- Tinner's hammer

- Upholstery hammer

- Welder's chipping hammer

Mechanically-powered Hammers

Mechanically-powered hammers often look quite different from the hand tools, but nevertheless most of them work on the same principle. They include:

Large mechanically-powered hammer

- Hammer drill, that combines a jackhammer-like mechanism with a drill

- High Frequency Impact Treatment hammer — for aftertreatment of weld transitions

- Jackhammer

- Steam hammer

- Trip hammer

In professional framing carpentry, the manual hammer has almost been completely replaced by the nail gun. In professional upholstery, its chief competitor is the staple gun.

Tools Used in Conjunction with Hammers

- Anvil

- Chisel

- Pipe drift (Blacksmithing - spreading a punched hole to proper size and/or shape)

- Star drill

- Punch

- Woodsplitting maul – can be hit with a sledgehammer for splitting wood.

- Woodsplitting wedge – hit with a sledgehammer for splitting wood.

Physics of Hammering

Hammer as A Force Amplifier

A hammer is basically a force amplifier that works by converting mechanical work into kinetic energy and back.

In the swing that precedes each blow, the hammer head stores a certain amount of kinetic energy—equal to the length D of the swing times the force f produced by the muscles of the arm and by gravity. When the hammer strikes, the head is stopped by an opposite force coming from the target, equal and opposite to the force applied by the head to the target. If the target is a hard and heavy object, or if it is resting on some sort of anvil, the head can travel only a very short distance d before stopping. Since the stopping force F times that distance must be equal to the head's kinetic energy, it follows that F is much greater than the original driving force f—roughly, by a factor D/d. In this way, great strength is not needed to produce a force strong enough to bend steel, or crack the hardest stone.

Effect of The Head's Mass

The amount of energy delivered to the target by the hammer-blow is equivalent to one half the mass of the head times the square of the head's speed at the time of impact $\left(E = \frac{mv^2}{2}\right)$. While the energy delivered to the target increases linearly with mass, it increases quadratically with the speed. High tech titanium heads are lighter and allow for longer handles, thus increasing velocity and delivering the same energy with less arm fatigue than that of a heavier steel head hammer. A titanium head has about 3% recoil energy and can result in greater efficiency and less fatigue when compared to a steel head with up to 30% recoil. Dead blow hammers use special rubber or steel shot to absorb recoil energy, rather than bouncing the hammer head after impact.

Effect of The Handle

The handle of the hammer helps in several ways. It keeps the user's hands away from the point of impact. It provides a broad area that is better-suited for gripping by the hand. Most importantly, it allows the user to maximize the speed of the head on each blow. The primary constraint on additional handle length is the lack of space to swing the hammer. This is why sledge hammers, largely used in open spaces, can have handles that are much longer than a standard carpenter's hammer. The second most important constraint is more subtle. Even without considering the effects of fatigue, the longer the handle, the harder it is to guide the head of the hammer to its target at full speed.

Most designs are a compromise between practicality and energy efficiency. With too long a handle, the hammer is inefficient because it delivers force to the wrong place, off-target. With too short a handle, the hammer is inefficient because it doesn't deliver enough force, requiring more blows to complete a given task. Modifications have also been made with respect to the effect of the hammer on the user. Handles made of shock-absorbing materials or varying angles attempt to make it easier for the user to continue to wield this age-old device, even as nail guns and other powered drivers encroach on its traditional field of use.

As hammers must be used in many circumstances, where the position of the person using them cannot be taken for granted, trade-offs are made for the sake of practicality. In areas where one has plenty of room, a long handle with a heavy head (like a sledge hammer) can deliver the maximum amount of energy to the target. It is not practical to use such a large hammer for all tasks, however, and thus the overall design has been modified repeatedly to achieve the optimum utility in a wide variety of situations.

Effect of Gravity

Gravity exerts a force on the hammer head. If hammering downwards, gravity increases the acceleration during the hammer stroke and increases the energy delivered with each blow. If hammering upwards, gravity reduces the acceleration during the hammer stroke and therefore reduces the energy delivered with each blow. Some hammering methods, such as traditional mechanical pile drivers, rely entirely on gravity for acceleration on the down stroke.

Ergonomics and Injury Risks

Both manual and powered hammers can cause peripheral neuropathy or a variety of other ailments when used improperly. Awkward handles can cause repetitive stress injury (RSI) to hand and arm joints, and uncontrolled shock waves from repeated impacts can injure nerves and the skeleton.

War Hammers

A war hammer is a late medieval weapon of war intended for close combat action.

Symbolic Hammers

The hammer, being one of the most used tools by *Homo sapiens*, has been used very much in symbols such as flags and heraldry. In the Middle Ages, it was used often in blacksmith guild logos, as well as in many family symbols. The hammer and pick is used as a symbol of mining.

A well known symbol with a hammer in it is the Hammer and Sickle, which was the symbol of the former Soviet Union and is very interlinked with communism and early socialism. The hammer in this symbol represents the industrial working class (and the sickle represents the agricultural working class). The hammer is used in some coat of arms in (former) socialist countries like East Germany. Similarly, the Hammer and Sword symbolizes Strasserism, a strand of National Socialism seeking to appeal to the working class.

The gavel, a small wooden mallet, is used to symbolize a mandate to preside over a meeting or judicial proceeding, and a graphic image of one is used as a symbol of legislative or judicial decision-making authority.

In Norse mythology, Thor, the god of thunder and lightning, wields a hammer named Mjölnir. Many artifacts of decorative hammers have been found, leading modern practitioners of this religion to often wear reproductions as a sign of their faith.

Judah Maccabee was nicknamed "The Hammer", possibly in recognition of his ferocity in battle. The name "Maccabee" may derive from the Aramaic *maqqaba*.

In American folkore, the hammer of John Henry represents the strength and endurance of a man.

The hammer in the song "If I Had a Hammer" represents a relentless message of justice broadcast across the land. The song became a symbol of the American Civil Rights movement.

Saw

A saw is a tool consisting of a tough blade, wire, or chain with a hard toothed edge. It is used to cut through material, most often wood. The cut is made by placing the toothed edge against the material and moving it forcefully forth and less forcefully back or continuously forward. This force may be applied by hand, or powered by steam, water, electricity or other power source. An abrasive saw has a powered circular blade designed to cut through metal.

Terminology

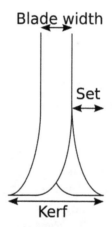

Diagram showing the teeth of a saw blade when looking front-on. The teeth protrude to the left and right, so that the saw cut (kerf) is wider than the blade width. The term *set* describes how much the teeth protrude. The kerf may be sometimes be wider than the set, depending on wobble and other factors.

- Abrasive saw: A saw that cuts with an abrasive disc or band, rather than a toothed blade.

- Back: the edge opposite the toothed edge.

- Fleam: The angle of the faces of the teeth relative to a line perpendicular to the face of the saw.

- Gullet: The valley between the points of the teeth.

- Heel: The end closest to the handle.

- Kerf: The width of a saw cut, which depends on several factors: the width of the saw blade; the set of the blade's teeth; the amount of wobble created during cutting; and the amount of material pulled out of the sides of the cut. Although the term "kerf" is often used informally, to refer simply to the thickness of the saw blade, or to the width of the set, this can be misleading, because blades with the same thickness and set may create different kerfs. For example, a too-thin blade can cause excessive wobble, creating a wider-than-expected kerf. The kerf created by a given blade can be changed by adjusting the set of its teeth with a tool called a saw tooth setter.

- Points per inch (25 mm): The most common measurement of the frequency of teeth on a saw blade. It is taken by setting the tip (or point) of one tooth at the zero point on a ruler, and then counting the number of points between the zero mark and the one-inch mark, inclusive (that is, including both the point at the zero mark and any point that lines up precisely with the one-inch mark). There is always one more point per inch than there are teeth per inch (e.g., a saw with 14 points per inch will have 13 teeth per inch, and a saw with 10 points per inch will have 9 teeth per inch). Some saws do not have the same number of teeth per inch throughout their entire length, but the vast majority do. Those with more teeth per inch at the toe are described as having incremental teeth, in order to make starting the saw cut easier.

- Rake: The angle of the front face of the tooth relative to a line perpendicular to the length of the saw. Teeth designed to cut with the grain (*ripping*) are generally steeper than teeth designed to cut across the grain (*crosscutting*)

- Set: The degree to which the teeth are bent out sideways away from the blade, usually in both directions. In most modern serrated saws, the teeth are set, so that the kerf (the width of the cut) will be wider than the blade itself. This allows the blade to move through the cut easily without *binding* (getting stuck). The set may be different depending on the kind of cut the saw is intended to make. For example, a rip saw has a tooth set that is similar to the angle used on a chisel, so that it rips or tears the material apart. A "flush-cutting saw" has no set on one side, so that the saw can be laid flat on a surface and cut along that surface without scratching it. The set of the blade's teeth can be adjusted with a tool called a saw set.

- Teeth: sharp protrusions along the cutting side of the saw.

- Teeth per inch: An alternative measurement of the frequency of teeth on a saw blade. Usually abbreviated TPI, as in, "A blade consisting of 18TPI." [Compare points per inch.]

- Toe: The end farthest from the handle.

- Toothed edge: the edge with the teeth (on some saws both edges are toothed).

- Web: a narrow saw blade held in a frame, worked either by hand or in a machine, sometimes with teeth on both edges

History

Saws were at first serrated materials such as flint, obsidian, sea shells and shark teeth.

Roman sawblades from Vindonissa approx. 3rd to 5th century AD

In ancient Egypt, open (unframed) saws made of copper are documented as early as the Early Dynastic Period, circa 3,100–2,686 BC. Saws have been used for cutting a variety of materials, including people. Models of saws have been found in many contexts throughout Egyptian history. Particularly useful are tomb wall illustrations of carpenters at work that show sizes and the use of different types. Egyptian saws were at first serrated, hardened copper which cut on both pull and push strokes. As the saw developed, teeth were raked to cut only on the pull stroke and set with the teeth projecting only on one side, rather than in the modern fashion with an alternating set. Saws were also made of bronze and later iron. In the Iron Age, frame saws were developed holding the thin blades in tension. The earliest known sawmill is the Roman Hierapolis sawmill from the third century AD and was for sawing stone.

Bronze-age saw blade from Akrotiri, late Cycladic period c. 17th century BC

According to Chinese legend, the saw was invented by Lu Ban. In Greek mythology, as recounted by Ovid, Talos, the nephew of Daedalus, invented the saw. In archeological reality, saws date back to prehistory and most probably evolved from Neolithic stone or bone tools. "[T]he identities of the axe, adz, chisel, and saw were clearly established more than 4,000 years ago."

Manufacture of Saws By Hand

Once mankind had learned how to use iron, this became the preferred material for saw blades of all kinds; some cultures learned how to harden the surface ("case hardening" or "steeling"), prolonging the blade's life and sharpness. Steel, made by quenching hot iron in water, was used as early as 1200 BC. By the end of the 17th century European manufacture centred on Germany (the Bergisches Land) and in London and the Midlands of England. Most blades were made of steel (iron carbonised and re-forged by different methods). In the mid 18th century a superior form of completely melted steel ("crucible cast") began to be made in Sheffield, England, and this rapidly became the preferred material, due to its hardness, ductility, springiness and ability to take a fine polish. A small saw industry survived in London and Birmingham, but by the 1820s the industry was growing rapidly and increasingly concentrated in Sheffield, which remained the largest centre of production, with over 50% of the nation's saw makers. The US industry began to overtake it in the last decades of the century, due to superior mechanisation, better marketing, a large domestic market, and the imposition of high tariffs on imports. Highly productive industries continued in Germany and France.

Early European saws were made from a heated sheet of iron or steel, produced by flattening by several men simultaneously hammering on an anvil <Barley ibid p11> After cooling, the teeth were punched out one at a time with a die, the size varying with the size of the saw. The teeth were sharpened with a triangular file of appropriate size, and set with a hammer or a wrest <Moxon, ibid>. By the mid 18th century rolling the metal was usual, the power for the rolls being supplied first by water, and increasingly by the early 19th century by steam engines. The industry gradually mechanized all the processes, including the important grinding the saw plate "thin to the back" by a fraction of an inch, which helped the saw to pass through the kerf without binding <Moxon, ibid, p95>. The use of steel added the need to harden and temper the saw plate, to grind it flat, to smith it by hand hammering and ensure the springiness and resistance to bending deformity, and finally to polish it <Barley ibid pp5–22>. Most hand saws are today entirely made without human intervention, with the steel plate supplied ready rolled to thickness and tensioned before being cut to shape by laser. The teeth are shaped and sharpened by grinding and are flame hardened to obviate (and actually prevent) sharpening once they have become blunt. A large measure of hand finishing remains to this day for quality saws by the very few specialist makers reproducing the 19th century designs.

Pit Saws

A pit saw was a two-man rip saw. In parts of early colonial North America, it was one of the principal tools used in shipyards and other industries where water-powered sawmills were not available. It was so-named because it was typically operated over a saw pit, either at ground level or on trestles across which logs that were to be cut into boards. The pit saw was "a strong steel cutting-plate, of great breadth, with large teeth, highly polished and thoroughly wrought, some eight or ten feet in length" with either a handle on each end or a frame saw. A pit-saw was also sometimes known as a whipsaw. It took 2-4 people to operate. A "pit-man" stood in the pit, a "top-man" stood outside the pit, and they worked together to make cuts, guide the saw, and raise it. Pit-saw workers were among the most highly paid laborers in early colonial North America.

Types of Saws

Hand Saws

Hand saws typically have a relatively thick blade to make them stiff enough to cut through material. (The pull stroke also reduces the amount of stiffness required.) Thin-bladed handsaws are made stiff enough either by holding them in tension in a frame, or by backing them with a folded strip of steel (formerly iron) or brass (on account of which the latter are called "back saws.") Some examples of hand saws are:

Rip sawing circa 1425 with a frame or sash saw on trestles rather than over a saw pit

- Artillery saw, Chain saw Portable link saw: a flexible chain saw up to four feet long, supplied to the military for clearing tree branches for gun sighting;

- Butcher's saw: for cutting bone; many different designs were common, including a large one for two men, known in the USA as a beef-splitter; most were frame saws, some backsaws;

- Crosscut saw: for cutting wood perpendicular to the grain;

- Docking saw: a large, heavy saw with an unbreakable metal handle of unique pattern, used for rough work

- Farmer's/Miner's saw: a strong saw with coarse teeth;

- Felloe saw;: the narrowest-bladed variety of pit saw, up to 7 feet long and able to work the sharp curves of cart wheel felloes; a slightly wider blade, equally long, was called a stave saw, for cutting the staves for wooden casks;

- Floorboard/flooring saw: a small saw, rarely with a back, and usually with the teeth continued onto the back at the toe for a short distance; used by house carpenters for cutting across a floor board with damaging its neighbour;

- Grafting/grafter/table saw; a hand saw with a tapering narrow blade from 6 to 30 inches long; the origins of the terms are obscure <Salaman, Dictionary, p420 and 433>

- Ice saw: either of pit saw design without a bottom tiller, or a large handsaw, always with very coarse teeth, for harvesting ice to be used away from source, or stored for use in warmer weather;

- Japanese saw or pull saw: a thin-bladed saw that cuts on the pull stroke, and with teeth of different design to European or American traditional forms;

- Keyhole/compass saw: a narrow-bladed saw, sharply tapered thin to the back to cut round curves, with one end fixed in a handle;

- Nest of saws: three or four interchangeable blades fitted to a handle with screws or quick-release nuts;

- One-man cross cut saw: a coarse-toothed saw of 30-60 inches length for rough or green timber; a second, turned, handle could be added at the heel or the toe for a second operator;

- Pad saw: a short narrow blade held in a wooden or metal handle (the pad);

- Panel saw: a lighter variety of handsaw, usually less than 24 inches long and having finer teeth;

- Plywood saw: a fine-toothed saw (to reduce tearing), for cutting plywood;

- Pruning saw: the commonest variety has a 12-28 inch blade, toothed on both edges, one tooth pattern being considerably coarser than the other;

- Rip saw: for cutting wood along the grain;

- Rule/combination saw; a handsaw with a measuring scale along the back and a handle making a 90degree square with the scaled edge;

- Salt saw: a short hand saw with a non-corroding zinc or copper blade, used for cutting a block of salt at a time when it was supplied to large kitchens in that form;

- Turkish/monkey saw: a small saw with a parallel-sided blade, designed to cut on the pull stroke;

- Two-man saw :a general term for a large crosscut saw or rip saw for cutting large logs or trees;

- Veneer saw: a two-edged saw with fine teeth for cutting veneer;

- Wire saw: a toothed or coarse cable or wire wrapped around the material and pulled back and forth.

Back Saws

"Back saws," so called because they have a thinner blade backed with steel or brass to maintain rigidity, are a subset of hand saws. Back saws have different names depending on the length of the blade; tenon saw is often used as a generic name for all the sizes of woodworking backsaw. Some examples are:

- Bead saw/gent's saw/jeweller's saw: a small backsaw with a turned wooden handle;

- Blitz saw: a small backsaw, for cutting wood or metal, with a hook at the toe for the thumb of the non-dominant hand;

- Carcase saw: a term used until the 20th century for backsaws with 10-14inch long blades;

- Dovetail saw: a backsaw with a blade of 6-10 inches length, for cutting intricate joints in cabinet making work;

- Electrician's saw: a very small backsaw used in the early 20th century on the wooden capping and casing in which electric wiring was run;

- Mitre-box saw: a saw with a blade 18-34 inches long, held in an adjustable frame (the mitre box) for making accurate crosscuts and mitres in a workplace;

- Sash saw: a backsaw of blade length 14-16 inches.

Frame Saws

A class of saws for cutting all types of material; they may be small or large and the frame may be wood or metal.

- Bow saw, Turning saw or Buck saw: a saw with a narrow blade held in tension in a frame; the blade can usually be rotated and may be toothed on both edges; it may be a rip or a crosscut, and was the preferred form of hand saw for continental European woodworkers until superseded by machines;

- Coping saw: a saw with a very narrow blade held in a metal frame in which it can usually be rotated, for cutting wood patterns;

- Felloe saw; a pit saw with a narrow tapering blade for sawing out the felloes of wooden cart wheels

- Fret saw: a saw with a very narrow blade which can be rotated, held in a deep metal frame, for cutting intricate wood patterns such as jigsaw puzzles;

- Girder saw: a large hack saw with a deep frame;

- Hacksaw/bow saw for iron: a fine-toothed blade held in a frame, for cutting metal, and other hard materials;

- Pit saw/sash saw/whip saw: large wooden framed saws for converting timber to lumber, with blades of various widths and lengths up to 10 feet; the timber is supported over a pit or raised on trestles; other designs are open bladed;

- Stave saw: a narrow tapering-bladed pit saw for sawing out staves for wooden casks;

- Surgeon's/surgical saw: for cutting bone during surgical procedures; some designs are framed, others have an open blade with a characteristic shape of the toe.

Mechanically Powered Saws

Circular-blade Saws

Circular wood-cutting saw at Maine State Museum in the capital city of Augusta, Maine

Reconstruction of the hydraulic saw by Leonardo da Vinci (*Codice Atlantico foglio 1078*) exposed at the Museo nazionale della scienza e della tecnologia Leonardo da Vinci, Milan.

- Circular saw: a saw with a circular blade which spins. Circular saws can be large for use in a mill or hand held up to 24" blades and different designs cut almost any kind of material including wood, stone, brick, plastic, etc.

- Table saw: a saw with a circular blade rising through a slot in a table. If it has a direct-drive blade small enough to set on a workbench, it is called a "work-bench" or "jobsite" saw. If set on steel legs, it is called a "contractor's saw." A heavier, more precise and powerful version, driven by several belts, with an enclosed base stand, is called a "cabinet saw." A newer version, combining the lighter-weight mechanism of a contractor's saw with the enclosed base stand of a cabinet saw, is called a "hybrid saw."

- Radial arm saw: a versatile machine, mainly for cross-cutting. The blade is pulled on a guide arm through a piece of wood that is held stationary on the saw's table.

- Rotary saw or "spiral-cut saw" or "RotoZip": for making accurate cuts, without using a pilot hole, in wallboard, plywood, and other thin materials.

- Electric miter saw or "chop saw," or "cut-off saw" or "power miter box": for making accurate cross cuts and miter cuts. The basic version has a circular blade fixed at a 90° angle to the vertical. A "compound miter saw" has a blade that can be adjusted to other angles. A "sliding compound miter saw" has a blade that can be pulled through the work, in an action similar to that of a radial-arm saw, which provides more capacity for cutting wider workpieces.

- Concrete saw: (usually powered by an internal combustion engine and fitted with a Diamond Blade) for cutting concrete or asphalt pavement.

- Pendulum saw or "swing saw": a saw hung on a swinging arm, for the rough cross cutting of wood in a sawmill and for cutting ice out of a frozen river.

- Abrasive saw: a circular or reciprocating saw-like tool with an abrasive disc rather than a toothed blade, commonly used for cutting very hard materials. As it does not have regularly shaped edges the abrasive saw is not a saw in technical terms.

- Hole saw: ring-shaped saw to attach to a power drill, used for cutting a circular hole in material.

Reciprocating Blade Saws

- Jigsaw or "saber saw" (US): narrow-bladed saw, for cutting irregular shapes. (Also an old term for what is now more commonly called a "scroll saw.")

- Reciprocating saw or "sabre saw" (UK and Australia): a saw with an "in-and-out" or "up-and-down" action similar to a jigsaw, but larger and more power-ful, and using a longer stroke with the blade parallel to the barrel. Hand-held versions, sometimes powered by compressed air, are for demolition work or for cutting pipe.

- Scroll saw: for making intricate curved cuts ("scrolls").

- Dragsaw: for bucking logs (used before the invention of the chainsaw).

- Frame saw or sash saw: A thin bladed rip-saw held in tension by a frame used both manually and in sawmills. Some whipsaws are frame saws and some have a heavy blade which does not need a frame called a mulay or muley saw.

- Sternal saw: for cutting through a patient's sternum during surgery.

- Ice saw: for ice cutting. Looks like a mulay saw but sharpened as a cross-cut saw.

Continuous Band

- Band saw: a ripsaw on a motor-driven continuous band. Portable sawmills are typically band saw mills.

Chainsaws

- Chainsaw: an engine-driven saw with teeth on a chain normally used as a cross-cut saw.

- Chainsaw mill: a chainsaw with a special saw chain and guide system for use as a rip-saw.

Types of Blades and Blade Cuts

Most blade teeth are made either of tool steel or carbide. Carbide is harder and holds a sharp edge much longer.

Band Saw Blade

A long band welded into a circle, with teeth on one side. Compared to a circu-lar-saw blade, it produces less waste because it is thinner, dissipates heat better because it is longer (so there is more blade to do the cutting, and is usually run at a slower speed.

Crosscut

In woodworking, a cut made at (or close to) a right angle to the direction of the wood grain of the workpiece. A crosscut saw is used to make this type of cut.

Rip Cut

In woodworking, a cut made parallel to the direction of the grain of the work-piece. A rip saw is used to make this type of cut.

Plytooth Blade

A circular saw blade with many small teeth, designed for cutting plywood with minimal splintering.

Dado Blade

A special type of circular saw blade used for making wide-grooved cuts in wood so that the edge of another piece of wood will fit into the groove to make a joint. Some dado blades can be adjusted to make different-width grooves. A "stacked" dado blade, consisting of chipper blades between two dado blades, can make different-width grooves by adding or removing chipper blades. An "adjustable" dado blade has a movable locking cam mechanism to adjust the degree to which the blade wobbles sideways, allowing continuously variable groove widths from the lower to upper design limits of the dado.

Strobe Saw Blade

A circular saw blade with special rakers/cutters to easily saw through green or uncured wood that tends to jam other kinds of saw blades.

Materials Used for Saws

There are several materials used in saws, with each of its own specifications.

Brass

Used only for the reinforcing folded strip along the back of backsaws, and to make the screws that in earlier times held the blade to the handle.

Iron

Used for blades and for the reinforcing strip on cheaper backsaws until super-seded by steel.

Zinc Used only for saws made to cut blocks of salt, as formerly used in kitchens

Copper Used as an alternative to zinc for salt-cutting saws

Steel

Used in almost every existing kind of saw. Because steel is cheap, easy to shape, and very strong, it has the right properties for most kind of saws.

Diamond

Fixed onto the saw blade's base to form diamond saw blades. As diamond is a superhard material, diamond saw blades can be used to cut hard brittle or abrasive materials, for example, stone, concrete, asphalt, bricks, ceramics, glass, semiconductor and gem stone. There are many methods used to fix the diamonds onto the blades' base and there are various kinds of diamond saw blades for different purposes. High speed steel (HSS): The whole saw blade is made of High Speed Steel (HSS). HSS saw blades are mainly used to cut steel, copper, aluminum and other metal materials. If high-strength steels (e.g., stainless steel) are to be cut, the blades made of cobalt HSS (e.g. M35, M42) should be used.

Tungsten carbide

Normally, there are two ways to use tungsten carbide to make saw blades:

1. Carbide-tipped saw blades: The saw blade's teeth are tipped (via welding) with small pieces of sharp tungsten carbide block. This type of blade is also called TCT (Tungsten Carbide-Tipped) saw blade. Carbide-tipped saw blades are widely used to cut wood, plywood, laminated board, plastic, grass, aluminum and some other metals.

2. Solid-carbide saw blades: The whole saw blade is made of tungsten carbide. Comparing with HSS saw blades, solid-carbide saw blades have higher hardness under high temperatures, and are more durable, but they also have a lower toughness.

Uses

A man recording the sound of a saw for sound effect purposes in the 1930s

- Saws are commonly used for cutting hard materials. They are used extensively in forestry, construction, demolition, medicine, and hunting.

- Musical saws are used as instruments to make music.

- Chainsaw carving is a flourishing modern art form. Special saws have been developed for the purpose.

- The production of lumber, lengths of squared wood for use in construction, begins with the felling of trees and the transportation of the logs to a sawmill.

Plainsawing: Lumber that will be used in structures is typically plainsawn (also called flatsawn), a method of dividing the log that produces the maximum yield of useful pieces and therefore the greatest economy.

Quarter sawing: This sawing method produces edge-grain or vertical grain lumber, in which annual growth rings run more consistently perpendicular to the pieces' wider faces.

Cutting Tool (Machining)

In the context of machining, a cutting tool or cutter is any tool that is used to remove material from the workpiece by means of shear deformation. Cutting may be accomplished by single-point or multipoint tools. Single-point tools are used in turning, shaping, planing and similar operations, and remove material by means of one cutting edge. Milling and drilling tools are often multipoint tools. Grinding tools are also multipoint tools. Each grain of abrasive functions as a microscopic single-point cutting edge (although of high negative rake angle), and shears a tiny chip.

Cutting tools must be made of a material harder than the material which is to be cut, and the tool must be able to withstand the heat generated in the metal-cutting process. Also, the tool must have a specific geometry, with clearance angles designed so that the cutting edge can contact the workpiece without the rest of the tool dragging on the workpiece surface. The angle of the cutting face is also important, as is the flute width, number of flutes or teeth, and margin size. In order to have a long working life, all of the above must be optimized, plus the speeds and feeds at which the tool is run.

Types

Linear cutting tools include tool bits (single-point cutting tools) and broaches. Rotary cutting tools include drill bits, countersinks and counterbores, taps and dies, milling cutters, reamers, and cold saw blades. Other cutting tools, such as bandsaw blades, hacksaw blades, and fly cutters, combine aspects of linear and rotary motion.

Cutting Tools with Inserts (Indexable Tools)

Cutting tools are often designed with inserts or replaceable tips (tipped tools). In these, the cutting edge consists of a separate piece of material, either brazed, welded or clamped on to the tool body. Common materials for tips include cemented carbide, polycrystalline diamond, and cubic boron nitride. Tools using inserts include milling cutters (endmills, fly cutters), tool bits, and saw blades.

Solid Cutting Tools

The typical tool for milling and drilling has no changeable insert. The cutting edge and the shank is one unit and built of the same material. Small tools cannot be designed with exchangeable inserts.

Holder

To use a cutting tool within a CNC machine there is a basic holder required to mount it on the machine's spindle or turret. For CNC milling machines, there are two types of holder. There are shank taper (SK) and hollow shank taper (HSK).

Tool Setup

The detailed instruction how to combine the tool assembly out of basic holder, tool and insert can be stored in a tool management solution.

Cutting Edge

The cutting edge of a cutting tool is a very important for the performance of the cutting process. The main features of the cutting edge are:

- form of the cutting edge: radius or waterfall or trumpete
- cutting edge angles (free angle and rake angle)
- form and size of the chamfers

The measurement of the cutting edge is performed using a tactile instrument or an instrument using focus variation.

Materials

To produce quality product, a cutting tool must have three characteristics:

- Hardness: hardness and strength at high temperatures.
- Toughness: so that tools do not chip or fracture.
- Wear resistance: having acceptable tool life before needing to be replaced.

Cutting tool materials can be divided into two main categories: stable and unstable.

Unstable materials (usually steels) are substances that start at a relatively low hardness point and are then heat treated to promote the growth of hard particles (usually carbides) inside the original matrix, which increases the overall hardness of the material at the expense of some its original toughness. Since heat is the mechanism to alter the structure of the substance and at the same time the cutting action produces a lot of heat, such substances are inherently unstable under machining conditions.

Stable materials (usually tungsten carbide) are substances that remain relatively stable under the heat produced by most machining conditions, as they don't attain their hardness through heat. They wear down due to abrasion, but generally don't change their properties much during use.

Most stable materials are hard enough to break before flexing, which makes them very fragile. To avoid chipping at the cutting edge, some tools made of such materials are finished with a sightly blunt edge, which results in higher cutting forces due to an increased shear area, however, tungsten carbide has the ability to attain a significantly sharper cutting edge than tooling steel for uses such as ultrasonic machining of composites. Fragility combined with high cutting forces results in most stable materials being unsuitable for use in anything but large, heavy and rigid machinery and fixtures.

Unstable materials, being generally softer and thus tougher, generally can stand a bit of flexing without breaking, which makes them much more suitable for unfavorable machining conditions, such as those encountered in hand tools and light machinery.

Tool material	Properties
Carbon tool steels	Unstable. Very inexpensive. Extremely sensitive to heat. Mostly obsolete in today's commercial machining, although it is still commonly found in non-intensive applications such as hobbyist or MRO machining, where economy-grade drill bits, taps and dies, hacksaw blades, and reamers are still usually made of it (because of its affordability). Hardness up to about HRC 65. Sharp cutting edges possible.
High speed steel (HSS)	Unstable. Inexpensive. Retains hardness at moderate temperatures. The most common cutting tool material used today. Used extensively on drill bits and taps. Hardness up to about HRC 67. Sharp cutting edges possible.
HSS cobalt	Unstable. Moderately expensive. The high cobalt versions of high speed steel are very resistant to heat and thus excellent for machining abrasive and/or work hardening materials such as titanium and stainless steel. Used extensively on milling cutters and drill bits. Hardness up to about HRC 70. Sharp cutting edges possible.
Cast cobalt alloys	Stable. Expensive. Somewhat fragile. Despite its stability it doesn't allow for high machining speed due to low hardness. Not used much. Hardness up to about HRC 65. Sharp cutting edges possible.
Cemented carbide	Stable. Moderately expensive. The most common material used in the industry today. It is offered in several "grades" containing different proportions of tungsten carbide and binder (usually cobalt). High resistance to abrasion. High solubility in iron requires the additions of tantalum carbide and niobium carbide for steel usage. Its main use is in turning tool bits although it is very common in milling cutters and saw blades. Hardness up to about HRA 93. Sharp edges generally not recommended.
Ceramics	Stable. Moderately inexpensive. Chemically inert and extremely resistant to heat, ceramics are usually desirable in high speed applications, the only drawback being their high fragility. Ceramics are considered unpredictable under unfavorable conditions. The most common ceramic materials are based on alumina (aluminium oxide), silicon nitride and silicon carbide. Used almost exclusively on turning tool bits. Hardness up to about HRC 93. Sharp cutting edges and positive rake angles are to be avoided.

Cermets	Stable. Moderately expensive. Another cemented material based on titanium carbide (TiC)or titanium carbonitride(TiCN). Binder is usually nickel. It provides higher abrasion resistance compared to tungsten carbide at the expense of some toughness. It is far more chemically inert than it too. Extremely high resistance to abrasion. Used primarily on turning tool bits although research is being carried on producing other cutting tools. Hardness up to about HRA 94. Sharp edges generally not recommended.
Cubic boron nitride (CBN)	Stable. Expensive. Being the second hardest substance known, it is also the second most fragile. It offers extremely high resistance to abrasion at the expense of much toughness. It is generally used in a machining process called "hard machining", which involves running the tool or the part fast enough to melt it before it touches the edge, softening it considerably. Used almost exclusively on turning tool bits. Hardness higher than HRC95. Sharp edges generally not recommended.
Diamond	Stable. Very expensive. The hardest substance known to date. Superior resistance to abrasion but also high chemical affinity to iron which results in being unsuitable for steel machining. It is used where abrasive materials would wear anything else. Extremely fragile. Used almost exclusively on turning tool bits although it can be used as a coating on many kinds of tools. Sharp edges generally not recommended.

Shaper

A shaper is a type of machine tool that uses linear relative motion between the work-piece and a single-point cutting tool to machine a linear toolpath. Its cut is analogous to that of a lathe, except that it is (archetypally) linear instead of helical.

Shaper tool slide, clapper box and cutting tool

Shaper with boring bar setup to allow cutting of internal features, such as keyways, or even shapes that might otherwise be cut with wire EDM.

A wood shaper is a similar woodworking tool, typically with a powered cutting head and manually fed workpiece, usually known simply as a *shaper* in North America and *spindle moulder* in the UK.

A metalworking shaper is somewhat analogous to a metalworking planer, with the cutter riding a ram that moves relative to a stationary workpiece, rather than the workpiece moving beneath the cutter. The ram is typically actuated by a mechanical crank inside the column, though hydraulically actuated shapers are increasingly used. Adding axes of motion to a shaper can yield helical toolpaths, as also done in helical planing.

Types

Shapers are mainly classified as standard, draw-cut, horizontal, universal, vertical, geared, crank, hydraulic, contour and traveling head, with a horizontal arrangement most common. Vertical shapers are generally fitted with a rotary table to enable curved surfaces to be machined (same idea as in helical planing). The vertical shaper is essentially the same thing as a slotter (slotting machine), although technically a distinction can be made if one defines a true vertical shaper as a machine whose slide can be moved from the vertical. A slotter is fixed in the vertical plane.

Operation

The workpiece mounts on a rigid, box-shaped table in front of the machine. The height of the table can be adjusted to suit this workpiece, and the table can traverse sideways underneath the reciprocating tool, which is mounted on the ram. Table motion may be controlled manually, but is usually advanced by an automatic feed mechanism acting on the feedscrew. The ram slides back and forth above the work. At the front end of the ram is a vertical tool slide that may be adjusted to either side of the vertical plane along the stroke axis. This tool-slide holds the *clapper box* and toolpost, from which the tool can be positioned to cut a straight, flat surface on the top of the workpiece. The tool-slide permits feeding the tool downwards to deepen a cut. This adjustability, coupled with the use of specialized cutters and toolholders, enable the operator to cut internal and external gear tooth.

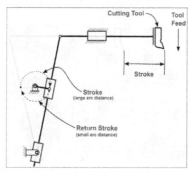

Shaper linkage. Note the drive arm revolves less for the return stroke than for the cutting stroke, resulting in a quicker return stroke and more powerful cutting stroke.

The ram is adjustable for stroke and, due to the geometry of the linkage, it moves faster on the return (non-cutting) stroke than on the forward, cutting stroke. This action is via a slotted link (or *Whitworth* link).

Uses

The most common use is to machine straight, flat surfaces, but with ingenuity and some accessories a wide range of work can be done. Other examples of its use are:

- Keyways in the boss of a pulley or gear can be machined without resorting to a dedicated broaching setup.

- Dovetail slides

- Internal splines and gear teeth.

- Keyway, spline, and gear tooth cutting in blind holes

- Cam drums with toolpaths of the type that in CNC milling terms would require 4- or 5-axis contouring or turn-mill cylindrical interpolation

- It is even possible to obviate wire EDM work in some cases. Starting from a drilled or cored hole, a shaper with a boring-bar type tool can cut internal features that don't lend themselves to milling or boring (such as irregularly shaped holes with tight corners).

- Smoothness of a rough surface

History

Roe (1916) credits James Nasmyth with the invention of the shaper in 1836. Shapers were very common in industrial production from the mid-19th century through the mid-20th. In current industrial practice, shapers have been largely superseded by other machine tools (especially of the CNC type), including milling machines, grinding machines, and broaching machines. But the basic function of a shaper is still sound; tooling for them is minimal and very cheap to reproduce; and they are simple and robust in construction, making their repair and upkeep easily achievable. Thus they are still popular in many machine shops, from jobbing shops or repair shops to tool and die shops, where only one or a few pieces are required to be produced and the alternative methods are cost- or tooling-intensive. They also have considerable retro appeal to many hobbyist machinists, who are happy to obtain a used shaper or, in some cases, even to build a new one from scratch.

Planer (Metalworking)

A planer is a type of metalworking machine tool that uses linear relative motion between the workpiece and a single-point cutting tool to cut the work piece. A planer is similar to a shaper, but larger, and with workpiece moving, whereas in a shaper the cutting tool moves.

A typical planer

- Watch video: Demonstration of metal planer on YouTube

Applications

Linear Planing

The most common applications of planers and shapers are linear-toolpath ones, such as:

- Generating accurate flat surfaces. (While not as precise as grinding, a planer can remove a tremendous amount of material in one pass with high accuracy.)

- Cutting slots (such as keyways).

- It is even possible to do work that might now be done by wire EDM in some cases. Starting from a drilled or cored hole, a planer with a boring-bar type tool can cut internal features that don't lend themselves to milling or boring (such as irregularly shaped holes with tight corners).

Helical planing

Although the archetypal toolpath of a planer is linear, helical cutting can be accomplished by coupling the table's linear motion to simultaneous rotation. The helical planing idea is similar to both helical milling and single-point screw cutting.

Current Usage

Planers and shapers are now obsolescent, because other machine tools (such as milling machines, broaching machines, and grinding machines) have mostly eclipsed them as the tools of choice for doing such work. However, they have not yet disappeared from the metalworking world. Planers are used by smaller tool and die shops within larger production facilities to maintain and repair large stamping dies and plastic injection molds. Additional uses include any other task where an abnormally large (usually in the range of 4'×8' or more) block of metal must be squared when a (quite massive) horizontal grinder or floor mill is unavailable, too expensive, or otherwise impractical in a given situation. As usual in the selection of machine tools, an old machine that is in hand, still works, and is long since paid-for has substantial cost advantage over a newer machine that would need to be purchased. This principle easily explains why "old-fashioned" techniques often have a long period of gradual obsolescence in industrial contexts, rather than a sharp drop-off of prevalence such as is seen in mass-consumer technology fashions.

Configurations and Sizes

There are two types of planers for metal: double-housing and open-side. The double-housing variety has vertical supports on both sides of its long bed; the open-side variety has a vertical support on only one side, allowing the workpiece to extend beyond the bed. Metal planers can vary in size from a table size of 30"×72" to 20'×62', and in weight from around 20,000 lbs to over 1,000,000 lbs.

History

Early planing ideas are known to have been underway in France in the 1750s. In the late 1810s, a variety of pioneers in various British shops (including James Fox, George Rennie, Matthew Murray, Joseph Clement, and Richard Roberts) developed the planer into what we today would call a machine tool. The exact details have been contentious and will probably never be known, because the development work being done in various shops was undocumented for various reasons (partially because of proprietary secrecy, and also simply because no one was taking down records for posterity). Roe (1916) provides a short chapter that tells the story as thoroughly as he was able to discover it.

Lathe

A lathe is a tool that rotates the workpiece on its axis to perform various operations such as cutting, sanding, knurling, drilling, or deformation, facing, turning, with tools that are applied to the workpiece to create an object with symmetry about an axis of rotation.

Lathes are used in woodturning, metalworking, metal spinning, thermal spraying, parts reclamation, and glass-working. Lathes can be used to shape pottery, the best-known design being the potter's wheel. Most suitably equipped metalworking lathes can also be used to produce most solids of revolution, plane surfaces and screw threads or helices. Ornamental lathes can produce three-dimensional solids of incredible complexity. The workpiece is usually held in place by either one or two *centers*, at least one of which can typically be moved horizontally to accommodate varying workpiece lengths. Other work-holding methods include clamping the work about the axis of rotation using a chuck or collet, or to a faceplate, using clamps or dogs.

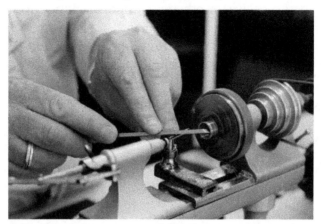

A watchmaker using a lathe to prepare a component cut from copper for a watch

Lathe, p. 1218.

A metalworking lathe from 1911, showing component parts:
a – bed
b – carriage (with cross-slide and toolpost)
c – headstock
d – back gear (other geartrain nearby drives leadscrew)
e – cone pulley for a belt drive from an external power source
f – faceplate mounted on spindle
g – tailstock
h – leadscrew

Examples of objects that can be produced on a lathe include candlestick holders, gun barrels, cue sticks, table legs, bowls, baseball bats, musical instruments (especially woodwind instruments), crankshafts, and camshafts.

History

The lathe is an ancient tool, dating at least to ancient Egypt and known to be used in Assyria and ancient Greece. The lathe was very important to the Industrial Revolution. It is known as the *mother of machine tools*, as it was the first machine tool that lead to the invention of other machine tools.

Craftsman Gregorio Vara working a chair leg on a lathe in Tenancingo, State of Mexico

The origin of turning dates to around 1300 BCE when the Ancient Egyptians first developed a two-person lathe. One person would turn the wood work piece with a rope while the other used a sharp tool to cut shapes in the wood. Ancient Rome improved the Egyptian design with the addition of a turning bow. In the Middle Ages a pedal replaced hand-operated turning, allowing a single person to rotate the piece while working with both hands. The pedal was usually connected to a pole, often a straight-grained sapling. The system today is called the "spring pole" lathe. Spring pole lathes were in common use into the early 20th century.

Exact drawing made with camera obscura of horizontal boring machine by Jan Verbruggen in Woolwich Royal Brass Foundry approx 1778 (drawing 47 out of set of 50 drawings)

An important early lathe in the UK was the horizontal boring machine that was installed in 1772 in the Royal Arsenal in Woolwich. It was horse-powered and allowed for the production of much more accurate and stronger cannon used with success in the American Revolutionary War in the late 18th century. One of the key characteristics of this machine was that the workpiece was turning as opposed to the tool, making it technically a lathe. Henry Maudslay who later developed many improvements to the lathe worked at the Royal Arsenal from 1783 being exposed to this machine in the Verbruggen workshop.

During the Industrial Revolution, mechanized power generated by water wheels or steam engines was transmitted to the lathe via line shafting, allowing faster and easier work. Metalworking lathes evolved into heavier machines with thicker, more rigid parts. Between the late 19th and mid-20th centuries, individual electric motors at each lathe replaced line shafting as the power source. Beginning in the 1950s, servomechanisms were applied to the control of lathes and other machine tools via numerical control, which often was coupled with computers to yield computerized numerical control (CNC). Today manually controlled and CNC lathes coexist in the manufacturing industries.

Description

Parts

A lathe may or may not have legs, which sit on the floor and elevate the lathe bed to a working height. A lathe may be small and sit on a workbench or table, not requiring a stand.

Almost all lathes have a bed, which is (almost always) a horizontal beam (although CNC lathes commonly have an inclined or vertical beam for a bed to ensure that swarf, or chips, falls free of the bed). Woodturning lathes specialized for turning large bowls often have no bed or tail stock, merely a free-standing headstock and a cantilevered tool rest.

At one end of the bed (almost always the left, as the operator faces the lathe) is a headstock. The headstock contains high-precision spinning bearings. Rotating within the bearings is a horizontal axle, with an axis parallel to the bed, called the spindle. Spindles are often hollow and have exterior threads and/or an interior Morse taper on the "inboard" (i.e., facing to the right / towards the bed) by which work-holding accessories may be mounted to the spindle. Spindles may also have exterior threads and/or an interior taper at their "outboard" (i.e., facing away from the bed) end, and/or may have a hand-wheel or other accessory mechanism on their outboard end. Spindles are powered and impart motion to the workpiece.

The spindle is driven either by foot power from a treadle and flywheel or by a belt or gear drive to a power source. In most modern lathes this power source is an integral electric motor, often either in the headstock, to the left of the headstock, or beneath the headstock, concealed in the stand.

In addition to the spindle and its bearings, the headstock often contains parts to convert the motor speed into various spindle speeds. Various types of speed-changing mechanism achieve this, from a cone pulley or step pulley, to a cone pulley with back gear (which is essentially a low range, similar in net effect to the two-speed rear of a truck), to an entire gear train similar to that of a manual-shift auto transmission. Some motors have electronic rheostat-type speed controls, which obviates cone pulleys or gears.

The counterpoint to the headstock is the tailstock, sometimes referred to as the loose head, as it can be positioned at any convenient point on the bed by sliding it to the required area. The tail-stock contains a barrel, which does not rotate, but can slide in and out parallel to the axis of the bed and directly in line with the headstock spindle. The barrel is hollow and usually contains a taper to facilitate the gripping of various types of tooling. Its most common uses are to hold a hardened steel center, which is used to support long thin shafts while turning, or to hold drill bits for drilling axial holes in the work piece. Many other uses are possible.

Metalworking lathes have a carriage (comprising a saddle and apron) topped with a cross-slide, which is a flat piece that sits crosswise on the bed and can be cranked at right angles to the bed. Sitting atop the cross slide is usually another slide called a compound rest, which provides 2 additional axes of motion, rotary and linear. Atop that sits a toolpost, which holds a cutting tool, which removes material from the workpiece. There may or may not be a leadscrew, which moves the cross-slide along the bed.

Woodturning and metal spinning lathes do not have cross-slides, but rather have banjos, which are flat pieces that sit crosswise on the bed. The position of a banjo can be adjusted by hand; no gearing is involved. Ascending vertically from the banjo is a toolpost, at the top of which is a horizontal tool-rest. In woodturning, hand tools are braced against the tool rest and levered into the workpiece. In metal spinning, the further pin ascends vertically from the tool rest and serves as a fulcrum against which tools may be levered into the workpiece.

Accessories

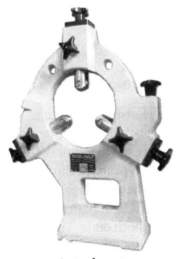

A steady rest

Unless a workpiece has a taper machined onto it which perfectly matches the internal taper in the spindle, or has threads which perfectly match the external threads on the spindle (two conditions which rarely exist), an accessory must be used to mount a workpiece to the spindle.

A workpiece may be bolted or screwed to a faceplate, a large, flat disk that mounts to the spindle. In the alternative, faceplate dogs may be used to secure the work to the faceplate.

A workpiece may be mounted on a mandrel, or circular work clamped in a three- or four-jaw chuck. For irregular shaped workpieces it is usual to use a four jaw (independent moving jaws) chuck. These holding devices mount directly to the lathe headstock spindle.

In precision work, and in some classes of repetition work, cylindrical workpieces are usually held in a collet inserted into the spindle and secured either by a draw-bar, or by a collet closing cap on the spindle. Suitable collets may also be used to mount square or hexagonal workpieces. In precision toolmaking work such collets are usually of the draw-in variety, where, as the collet is tightened, the workpiece moves slightly back into the headstock, whereas for most repetition work the dead length variety is preferred, as this ensures that the position of the workpiece does not move as the collet is tightened.

A soft workpiece (e.g., wood) may be pinched between centers by using a spur drive at the headstock, which bites into the wood and imparts torque to it.

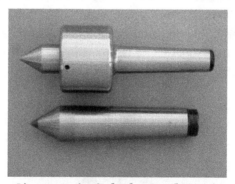

Live center (top); dead center (bottom)

A soft dead center is used in the headstock spindle as the work rotates with the centre. Because the centre is soft it can be trued in place before use. The included angle is 60°. Traditionally, a hard dead center is used together with suitable lubricant in the tailstock to support the workpiece. In modern practice the dead center is frequently replaced by a live center, as it turns freely with the workpiece — usually on ball bearings — reducing the frictional heat, especially important at high speeds. When clear facing a long length of material it must be supported at both ends. This can be achieved by the use of a traveling or fixed steady. If a steady is not available, the end face being worked on may be supported by a dead (stationary) half center. A half center has a flat surface machined across a broad section of half of its diameter at the pointed end. A small section of the tip of the dead center is retained to ensure concentricity. Lubrication must be applied at this point of contact and tail stock pressure reduced. A lathe carrier or lathe dog may also be employed when turning between two centers.

In woodturning, one variation of a live center is a cup center, which is a cone of metal surrounded by an annular ring of metal that decreases the chances of the workpiece splitting.

A circular metal plate with even spaced holes around the periphery, mounted to the spindle, is called an "index plate". It can be used to rotate the spindle to a precise angle, then lock it in place, facilitating repeated auxiliary operations done to the workpiece.

Other accessories, including items such as taper turning attachments, knurling tools, vertical slides, fixed and traveling steadies, etc., increase the versatility of a lathe and the range of work it may perform.

Modes of Use

When a workpiece is fixed between the headstock and the tail-stock, it is said to be "between centers". When a workpiece is supported at both ends, it is more stable, and more force may be applied to the workpiece, via tools, at a right angle to the axis of rotation, without fear that the workpiece may break loose.

When a workpiece is fixed only to the spindle at the headstock end, the work is said to be "face work". When a workpiece is supported in this manner, less force may be applied to the workpiece, via tools, at a right angle to the axis of rotation, lest the workpiece rip free. Thus, most work must be done axially, towards the headstock, or at right angles, but gently.

When a workpiece is mounted with a certain axis of rotation, worked, then remounted with a new axis of rotation, this is referred to as "eccentric turning" or "multi-axis turning". The result is that various cross sections of the workpiece are rotationally symmetric, but the workpiece as a whole is not rotationally symmetric. This technique is used for camshafts, various types of chair legs.

Varieties

The smallest lathes are "jewelers lathes" or "watchmaker lathes", which are small enough that they may be held in one hand. The workpieces machined on a jeweler's lathe are metal. Jeweler's lathes can be used with hand-held "graver" tools or with compound rests that attach to the lathe bed. Graver tools are generally supported by a T-rest, not fixed to a cross slide or compound rest. The work is usually held in a collet. Common spindle bore sizes are 6 mm, 8 mm and 10 mm. The term W/W refers to the Webster/Whitcomb collet and lathe, invented by the American Watch Tool Company of Waltham, Massachusetts. Most lathes commonly referred to as watchmakers lathes are of this design. In 1909, the American Watch Tool company introduced the Magnus type collet (a 10-mm body size collet) using a lathe of the same basic design, the Webster/Whitcomb Magnus. (F.W.Derbyshire, Inc. retains the trade names Webster/Whitcomb and Magnus and still produces these collets.) Two bed patterns are common: the WW (Webster Whitcomb) bed, a truncated triangular prism (found only on 8 and 10 mm watchmakers' lathes); and the continental D-style bar bed (used on both 6 mm and 8 mm lathes by firms such as Lorch and Star). Other bed designs have been used, such a triangular prism on some Boley 6.5 mm lathes, and a V-edged bed on IME's 8 mm lathes.

Smaller metalworking lathes that are larger than jewelers' lathes and can sit on a bench or table, but offer such features as tool holders and a screw-cutting gear train are called hobby lathes, and larger versions, "bench lathes". Even larger lathes offering similar features for producing or modifying individual parts are called "engine lathes". Lathes of these types do not have additional integral features for repetitive production, but rather are used for individual part production or modification as the primary role.

Lathes of this size that are designed for mass manufacture, but not offering the versatile screw-cutting capabilities of the engine or bench lathe, are referred to as "second operation" lathes.

Lathes with a very large spindle bore and a chuck on both ends of the spindle are called "oil field lathes".

Fully automatic mechanical lathes, employing cams and gear trains for controlled movement, are called screw machines.

Lathes that are controlled by a computer are CNC lathes.

Lathes with the spindle mounted in a vertical configuration, instead of horizontal configuration, are called vertical lathes or vertical boring machines. They are used where very large diameters must be turned, and the workpiece (comparatively) is not very long.

A lathe with a cylindrical tail-stock that can rotate around a vertical axis, so as to present different tools towards the headstock (and the workpiece) are turret lathes.

A lathe equipped with indexing plates, profile cutters, spiral or helical guides, etc., so as to enable ornamental turning is an ornamental lathe.

Various combinations are possible: for example, a vertical lathe can have CNC capabilities as well (such as a CNC VTL).

Lathes can be combined with other machine tools, such as a drill press or vertical milling machine. These are usually referred to as combination lathes.

Major Categories

Woodworking Lathes

Woodworking lathes are the oldest variety. All other varieties are descended from these simple lathes. An adjustable horizontal metal rail – the tool rest – between the material and the operator accommodates the positioning of shaping tools, which are usually hand-held. After shaping, it is common practice to press and slide sandpaper against the still-spinning object to smooth the surface made with the metal shaping tools. The tool rest is usually removed during sanding, as it may be unsafe to have the operators hands between it and the spinning wood.

A modern woodworking lathe

Many woodworking lathes can also be used for making bowls and plates. The bowl or plate needs only to be held at the bottom by one side of the lathe. It is usually attached to a metal face plate attached to the spindle. With many lathes, this operation happens on the left side of the headstock, where are no rails and therefore more clearance. In this configuration, the piece can be shaped inside and out. A specific curved tool rest may be used to support tools while shaping the inside. Further detail can be found on the woodturning page.

Most woodworking lathes are designed to be operated at a speed of between 200 and 1,400 revolutions per minute, with slightly over 1,000 rpm considered optimal for most such work, and with larger workpieces requiring lower speeds.

Duplicating Lathes

Water-powered Blanchard lathe used for duplicating gun stocks from 1850's. Harpers Ferry Armory.

One type of specialized lathe is duplicating or copying lathe also known as Blanchard lathe after its inventor Thomas Blanchard. This type of lathe was able to create shapes identical to a standard pattern and it revolutionized the process of gun stock making in 1820's when it was invented.

Patternmaker's Lathes

Used to make a pattern for foundries, often from wood, but also plastics. A pattern-maker's lathe looks like a heavy wood lathe, often with a turret and either a leadscrew or a rack and pinion to manually position the turret. The turret is used to accurately cut straight lines. They often have a provision to turn very large parts on the other end of the headstock, using a free-standing toolrest. Another way of turning large parts is a sliding bed, which can slide away from the headstock and thus open up a gap in front of the headstock for large parts.

Patternmaker's double lathe (Carpentry and Joinery, 1925)

Metalworking Lathes

In a metalworking lathe, metal is removed from the workpiece using a hardened cutting tool, which is usually fixed to a solid moveable mounting, either a tool-post or a turret, which is then moved against the workpiece using handwheels and/or computer-controlled motors. These cutting tools come in a wide range of sizes and shapes, depending upon their application. Some common styles are diamond, round, square and triangular.

A CNC metalworking lathe

The tool-post is operated by lead-screws that can accurately position the tool in a variety of planes. The tool-post may be driven manually or automatically to produce the roughing and finishing cuts required to *turn* the workpiece to the desired shape and dimensions, or for cutting threads, worm gears, etc. Cutting fluid may also be pumped to the cutting site to provide cooling, lubrication and clearing of swarf from the workpiece. Some lathes may be operated under control of a computer for mass production of parts.

Manually controlled metalworking lathes are commonly provided with a variable-ratio gear-train to drive the main lead-screw. This enables different thread pitches to be cut. On some older lathes or more affordable new lathes, the gear trains are changed by swapping gears with various numbers of teeth onto or off of the shafts, while more modern or expensive manually controlled lathes have a quick-change box to provide commonly used ratios by the operation of a lever. CNC lathes use computers and servomechanisms to regulate the rates of movement.

On manually controlled lathes, the thread pitches that can be cut are, in some ways, determined by the pitch of the lead-screw: A lathe with a metric lead-screw will readily cut metric threads (including BA), while one with an imperial lead-screw will readily cut imperial-unit-based threads such as BSW or UTS (UNF, UNC). This limitation is not insurmountable, because a 127-tooth gear, called a transposing gear, is used to translate between metric and inch thread pitches. However, this is optional equipment that many lathe owners do not own. It is also a larger change-wheel than the others, and on some lathes may be larger than the change-wheel mounting banjo is capable of mounting.

The workpiece may be supported between a pair of points called centres, or it may be bolted to a faceplate or held in a chuck. A chuck has movable jaws that can grip the workpiece securely.

There are some effects on material properties when using a metalworking lathe. There are few chemical or physical effects, but there are many mechanical effects, which include residual stress, micro-cracks, work-hardening, and tempering in hardened materials.

Cue Lathes

Cue lathes function similarly to turning and spinning lathes, allowing a perfectly radially-symmetrical cut for billiard cues. They can also be used to refinish cues that have been worn over the years.

Glass-working Lathes

Glass-working lathes are similar in design to other lathes, but differ markedly in how the workpiece is modified. Glass-working lathes slowly rotate a hollow glass vessel over a fixed- or variable-temperature flame. The source of the flame may be either hand-held or mounted to a banjo/cross-slide that can be moved along the lathe bed. The flame serves to soften the glass being worked, so that the glass in a specific area of the workpiece becomes ductile and subject to forming either by inflation ("glassblowing") or by deformation with a heat-resistant tool. Such lathes usually have two head-stocks with chucks holding the work, arranged so that they both rotate together in unison. Air can be introduced through the headstock chuck spindle for glassblowing. The tools to deform the glass and tubes to blow (inflate) the glass are usually handheld.

In diamond turning, a computer-controlled lathe with a diamond-tipped tool is used to make precision optical surfaces in glass or other optical materials. Unlike conventional optical grinding, complex aspheric surfaces can be machined easily. Instead of the dovetailed ways used on the tool slide of a metal-turning lathe, the ways typically float on air bearings, and the position of the tool is measured by optical interferometry to achieve the necessary standard of precision for optical work. The finished work piece usually requires a small amount of subsequent polishing by conventional techniques to achieve a finished surface suitably smooth for use in a lens, but the rough grinding time is significantly reduced for complex lenses.

Metal-spinning Lathes

In metal spinning, a disk of sheet metal is held perpendicularly to the main axis of the lathe, and tools with polished tips (*spoons*) or roller tips are hand-held, but levered by hand against fixed posts, to develop pressure that deforms the spinning sheet of metal.

Metal-spinning lathes are almost as simple as wood-turning lathes. Typically, metal spinning requires a mandrel, usually made from wood, which serves as the template onto which the workpiece is formed (asymmetric shapes can be made, but it is a very advanced technique). For example, to make a sheet metal bowl, a solid block of wood in the shape of the bowl is required; similarly, to make a vase, a solid template of the vase is required.

Given the advent of high-speed, high-pressure, industrial die forming, metal spinning is less common now than it once was, but still a valuable technique for producing one-off prototypes or small batches, where die forming would be uneconomical.

Ornamental Turning Lathes

The ornamental turning lathe was developed around the same time as the industrial screw-cutting lathe in the nineteenth century. It was used not for making practical objects, but for decorative work – *ornamental turning*. By using accessories such as the horizontal and vertical cutting frames, eccentric chuck and elliptical chuck, solids of extraordinary complexity may be produced by various generative procedures.

A special-purpose lathe, the Rose engine lathe, is also used for ornamental turning, in particular for engine turning, typically in precious metals, for example to decorate pocket-watch cases. As well as a wide range of accessories, these lathes usually have complex dividing arrangements to allow the exact rotation of the mandrel. Cutting is usually carried out by rotating cutters, rather than directly by the rotation of the work itself. Because of the difficulty of polishing such work, the materials turned, such as wood or ivory, are usually quite soft, and the cutter has to be exceptionally sharp. The finest ornamental lathes are generally considered to be those made by Holtzapffel around the turn of the 19th century.

Reducing Lathe

Many types of lathes can be equipped with accessory components to allow them to reproduce an item: the original item is mounted on one spindle, the blank is mounted on another, and as both turn in synchronized manner, one end of an arm "reads" the original and the other end of the arm "carves" the duplicate.

A reduction lathe is a specialized lathe that is designed with this feature and incorporates a mechanism similar to a pantograph, so that when the "reading" end of the arm reads a detail that measures one inch (for example), the cutting end of the arm creates an analogous detail that is (for example) one quarter of an inch (a 4:1 reduction, although given appropriate machinery and appropriate settings, any reduction ratio is possible).

Reducing lathes are used in coin-making, where a plaster original (or an epoxy master made from the plaster original, or a copper-shelled master made from the plaster original, etc.) is duplicated and reduced on the reducing lathe, generating a master die.

Rotary Lathes

A lathe in which softwood, like spruce or pine, or hardwood, like birch, logs are turned against a very sharp blade and peeled off in one continuous or semi-continuous roll. Invented by Immanuel Nobel (father of the more famous Alfred Nobel). The first such lathes were set up in the United States in the mid-19th century. The product is called wood veneer and it is used for making plywood and as a cosmetic surface veneer on some grades of chipboard.

Watchmaker's Lathes

Watchmakers lathes are delicate but precise metalworking lathes, usually without provision for screwcutting, and are still used by horologists for work such as the turning of balance staffs. A handheld tool called a graver is often used in preference to a slide-mounted tool. The original watchmaker's turns was a simple dead-center lathe with a moveable rest and two loose head-stocks. The workpiece would be rotated by a bow, typically of horsehair, wrapped around it.

Transcription, or Recording, Lathes

Transcription or recording lathes are used to make grooves on a surface for recording sounds. These were used in creating sound grooves on wax cylinders and then on flat recording discs. Originally the cutting lathes were driven by sound vibrations through a horn and then later driven by electric current, when microphones were used in recording. Many of these were professional models, but there were some used for home recording and were popular before the advent of home tape recording.

Performance Evaluation

National and international standards are used to standardize the definitions, environmental requirements, and test methods used for the performance evaluation of lathes. Selection of the standard to be used is an agreement between the supplier and the user and has some significance in the design of the lathe. In the United States, ASME has developed the B5.57 Standard entitled "Methods for Performance Evaluation of Computer Numerically Controlled Lathes and Turning Centers", which establishes requirements and methods for specifying and testing the performance of CNC lathes and turning centers.

Four-slide

A four-slide, also known as a multislide, multi-slide, or four-way, is a metalworking machine tool used in the high-volume manufacture of small stamped components from bar or wire stock. The press is most simply described as a horizontal stamping press that uses cams to control tools. The machine is used for progressive or transfer stamping operations.

A four-slide machine

The same four-slide machine from another angle

Design

A four-slide is quite different from most other presses. The key of the machine is its moving slides that have tools attached, which strike the workpiece together or in sequence to form it. These slides are driven by four shafts that outline the machine. The shafts are connected by bevel gears so that one shaft is driven by an electric motor, and then that shaft's motion drives the other three shafts. Each shaft then has cams which drive the slides, usually of a split-type. This shafting arrangement allows the workpiece to be worked for four sides, which makes this machine extremely versatile. A hole near the center of the machine is provided to expel the completed workpiece.

Advantages and Disadvantages

The greatest advantage of the four-slide machine is its ability to complete all of the operations required to form the workpiece from start to finish. Moreover, it can handle certain parts that transfer or progressive dies cannot, because it can manipulate from four axes. Due to this flexibility it reduces the cost of the finished part because it requires less machines, setups, and handling. Also, because only one machine is required, less space is required for any given workpiece. As compared to standard stamping presses the tooling is usually inexpensive, due to the simplicity of the tools. A four-slide can usually produce 20,000 to 70,000 finished parts per 16-hour shift, depending on the number of operations per part; this speed usually results in a lower cost per part.

The biggest disadvantage is its size constraints. The largest machines can handle stock up to 3 in (76 mm) wide, 12.5 in (320 mm) long, and $\frac{3}{32}$ in (2.4 mm) thick. For wires the limit is $\frac{1}{8}$ in (3.175 mm). Other limits are the travel on the slides, which maxes out at $\frac{3}{4}$ in (19.05 mm), and the throw of the forming cams, which is between $\frac{7}{8}$ and 2 in (22 and 51 mm). The machine is also limited to only shearing and bending operations. Extrusion and upsetting operations are impractical because it hinders the movement of the workpiece to the next station. Drawing and stretching require too much tonnage and the mechanisms required for the operations are space prohibitive. Finally, this machine is only feasible to use on high volume parts because of the long lead time required to set up the tooling.

Materials

The material stock used in four-slides is usually limited by its formability and not the machine capabilities. Usually the forming characteristics and bending radii are the most limiting factors. The most commonly used materials are:

- Low-carbon cold rolled steel

- Spheroidized cold rolled spring steel

- Type 300 and 400 stainless spring steels

- Copper alloys

- Beryllium-copper alloys

Use

Items that are commonly produced on this machine include: automotive stampings, hinges, links, clips, and razor blades.

References

- Jacques W. Delleur (12 December 2010). The Handbook of Groundwater Engineering, Second Edition. Taylor & Francis. p. 7 in chapter 2. ISBN 978-0-8493-4316-2.

- Geng Ruilun (1 October 1997). Guo Huadong, ed. New Technology for Geosciences: Proceedings of the 30th International Geological Congress. VSP. p. 225. ISBN 978-90-6764-265-1.

- Nagyszalanczy, Sandor (2001). Power Tools: An Electrifying Celebration and Grounded Guide. Newtown, CT: The Taunton Press. ISBN 978-1-56158-427-7.

- Johnson, Roland (2010). Complete Illustrated guide to band saws (PDF). the Taunton Press. p. 6. ISBN 978-1-60085-096-7.

- Gang Zhao (1986). Man and land in Chinese history: an economic analysis, p. 65. Stanford University Press. pp. 65–. ISBN 978-0-8047-1271-2. Retrieved 8 April 2011.

- Broadman & Holman Publishers (15 September 2006). Holman Illustrated Study Bible-HCSB. B&H Publishing Group. pp. 1413–. ISBN 978-1-58640-275-4. Retrieved 8 April 2011.

- Frances Welsh (4 March 2008). Tutankhamun's Egypt. Osprey Publishing. pp. 7–. ISBN 978-0-7478-0665-3. Retrieved 19 April 2011.

- Martin Brennan (1980). The Boyne Valley vision. Dolmen Press. ISBN 978-0-85105-362-2. Retrieved 22 April 2011.

- Heggie, Douglas C. (1981). Megalithic Science: Ancient Mathematics and Astronomy in Northwest Europe. Thames and Hudson. p. 58. ISBN 0-500-05036-8.

- Donald Preziosi (1983). Minoan architectural design: formation and signification, p. 498. Mouton. ISBN 978-90-279-3409-3. Retrieved 20 April 2011.

- Dominique Collon (1975). The seal impressions from Tell Atchana/Alalakh. Butzon & Bercker. ISBN 978-3-7887-0469-8. Retrieved 22 April 2011.

- Amélie Kuhrt (1995). The ancient Near East, c. 3000-330 BC. Routledge. pp. 111–. ISBN 978-0-415-16763-5. Retrieved 8 April 2011.

- Lucinda Dirven (1999). The Palmyrenes of Dura-Europos: a study of religious interaction in Roman Syria p. 329. BRILL. pp. 329–. ISBN 978-90-04-11589-7. Retrieved 8 April 2011.

- Maarten Jozef Vermaseren; International Association for the History of Religions. Dutch Section (1979). Studies in Hellenistic religions. Brill Archive. pp. 199–. ISBN 978-90-04-05885-9. Retrieved 22 April 2011.

- Edward Washburn Hopkins (10 January 2007). The Religions of India. Echo Library. pp. 52–. ISBN 978-1-4068-1329-6. Retrieved 24 April 2011.

- W. R. Lethaby (December 2005). Architecture, Mysticism and Myth. Cosimo, Inc. pp. 16–. ISBN 978-1-59605-380-9. Retrieved 24 April 2011.

- Stephenson, David A.; Agapiou, John S. (1997), Metal cutting theory and practice, Marcel Dekker, p. 164, ISBN 978-0-8247-9579-5.

- Parker, Dana T. Building Victory: Aircraft Manufacturing in the Los Angeles Area in World War II, p. 73, Cypress, CA, 2013. ISBN 978-0-9897906-0-4.

- Ernie Conover (2000), Turn a Bowl with Ernie Conover: Getting Great Results the First Time Around, Taunton, p. 16, ISBN 978-1-56158-293-8.

Machining: Processes and Techniques

The processes and techniques used in machining are milling, grinding, abrasive machining, abrasive jet machining, productivity improving technologies, rapid prototyping and electrical discharge machining. Milling is the process that is used as a cutter in order to remove materials from any work piece. The aspects elucidated in this chapter are of vital importance, and provides a better understanding of machining.

Milling (Machining)

Milling is the machining process of using rotary cutters to remove material from a workpiece by advancing (or *feeding*) in a direction at an angle with the axis of the tool. It covers a wide variety of different operations and machines, on scales from small individual parts to large, heavy-duty gang milling operations. It is one of the most commonly used processes in industry and machine shops today for machining parts to precise sizes and shapes.

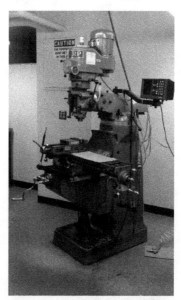

Full view of a Bridgeport clone.

Milling can be done with a wide range of machine tools. The original class of machine tools for milling was the milling machine (often called a mill). After the advent of computer numerical control (CNC), milling machines evolved into machining centers (milling machines with automatic tool changers, tool magazines or carousels, CNC control, coolant systems, and enclosures), generally classified as vertical machining centers (VMCs) and horizontal machining centers (HMCs). The integration of milling into turning environments and of turning into milling environments, begun with live tooling for lathes and the occasional use of mills for turning operations, led to a new class of machine tools, multitasking machines (MTMs), which are purpose-built to provide for a default machining strategy of using any combination of milling and turning within the same work envelope.

Process

Milling is a cutting process that uses a milling cutter to remove material from the surface of a workpiece. The milling cutter is a rotary cutting tool, often with multiple cutting points. As opposed to drilling, where the tool is advanced along its rotation axis, the cutter in milling is usually moved perpendicular to its axis so that cutting occurs on the circumference of the cutter. As the milling cutter enters the workpiece, the cutting edges (flutes or teeth) of the tool repeatedly cut into and exit from the material, shaving off chips (swarf) from the workpiece with each pass. The cutting action is shear deformation; material is pushed off the workpiece in tiny clumps that hang together to a greater or lesser extent (depending on the material) to form chips. This makes metal cutting somewhat different (in its mechanics) from slicing softer materials with a blade.

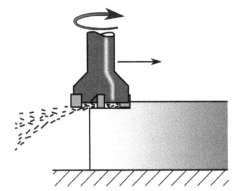
Face milling process (cutter rotation axis is vertical)

The milling process removes material by performing many separate, small cuts. This is accomplished by using a cutter with many teeth, spinning the cutter at high speed, or advancing the material through the cutter slowly; most often it is some combination of these three approaches. The speeds and feeds used are varied to suit a combination of variables. The speed at which the piece advances through the cutter is called feed rate, or just feed; it is most often measured in length of material per full revolution of the cutter.

There are two major classes of milling process:

- In face milling, the cutting action occurs primarily at the end corners of the milling cutter. Face milling is used to cut flat surfaces (faces) into the workpiece, or to cut flat-bottomed cavities.

- In peripheral milling, the cutting action occurs primarily along the circumference of the cutter, so that the cross section of the milled surface ends up receiving the shape of the cutter. In this case the blades of the cutter can be seen as scooping out material from the work piece. Peripheral milling is well suited to the cutting of deep slots, threads, and gear teeth.

Milling Cutters

Many different types of cutting tools are used in the milling process. Milling cutters such as endmills may have cutting surfaces across their entire end surface, so that they can be drilled into the workpiece (plunging). Milling cutters may also have extended cutting surfaces on their sides to allow for peripheral milling. Tools optimized for face milling tend to have only small cutters at their end corners.

The cutting surfaces of a milling cutter are generally made of a hard and temperature-resistant material, so that they wear slowly. A low cost cutter may have surfaces made of high speed steel. More expensive but slower-wearing materials include cemented carbide. Thin film coatings may be applied to decrease friction or further increase hardness.

They are cutting tools typically used in milling machines or machining centres to perform milling operations (and occasionally in other machine tools). They remove material by their movement within the machine (e.g., a ball nose mill) or directly from the cutter's shape (e.g., a form tool such as a hobbing cutter).

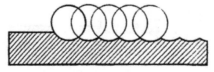

A diagram of revolution ridges on a surface milled by the side of the cutter, showing the position of the cutter for each cutting pass and how it corresponds with the ridges (cutter rotation axis is perpendicular to image plane)

As material passes through the cutting area of a milling machine, the blades of the cutter take swarfs of material at regular intervals. Surfaces cut by the side of the cutter (as in peripheral milling) therefore always contain regular ridges. The distance between ridges and the height of the ridges depend on the feed rate, number of cutting surfaces, the cutter diameter. With a narrow cutter and rapid feed rate, these revolution ridges can be significant variations in the surface finish.

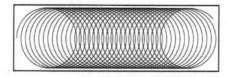

Trochoidal marks, characteristic of face milling.

The face milling process can in principle produce very flat surfaces. However, in practice the result always shows visible trochoidal marks following the motion of points on the cutter's end face. These revolution marks give the characteristic finish of a face milled surface. Revolution marks can have significant roughness depending on factors such as flatness of the cutter's end face and the degree of perpendicularity between the cutter's rotation axis and feed direction. Often a final pass with a slow feed rate is used to improve the surface finish after the bulk of the material has been removed.. In a precise face milling operation, the revolution marks will only be microscopic scratches due to imperfections in the cutting edge.

Gang Milling

Gang milling refers to the use of two or more milling cutters mounted on the same arbor (that is, ganged) in a horizontal-milling setup. All of the cutters may perform the same type of operation, or each cutter may perform a different type of operation. For example, if several workpieces need a slot, a flat surface, and an angular groove, a good method to cut these (within a non-CNC context) would be gang milling. All the completed workpieces would be the same, and milling time per piece would be minimized.

Heavy gang milling of milling machine tables

Gang milling was especially important before the CNC era, because for duplicate part production, it was a substantial efficiency improvement over manual-milling one feature at an operation, then changing machines (or changing setup of the same machine) to cut the next op. Today, CNC mills with automatic tool change and 4- or 5-axis control obviate gang-milling practice to a large extent.

Equipment

Milling is performed with a milling cutter in various forms, held in a collett or similar which, in turn, is held in the spindle of a milling machine.

Types and Nomenclature

Mill orientation is the primary classification for milling machines. The two basic configurations are vertical and horizontal. However, there are alternative classifications according to method of control, size, purpose and power source.

Mill Orientation

Vertical mill

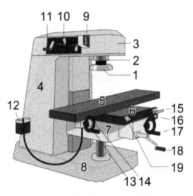

Vertical milling machine. 1: milling cutter 2: spindle 3: top slide or overarm 4: column 5: table 6: Y-axis slide 7: knee 8: base

In the vertical mill the spindle axis is vertically oriented. Milling cutters are held in the spindle and rotate on its axis. The spindle can generally be extended (or the table can be raised/lowered, giving the same effect), allowing plunge cuts and drilling. There are two subcategories of vertical mills: the bed mill and the turret mill.

- A turret mill has a stationary spindle and the table is moved both perpendicular and parallel to the spindle axis to accomplish cutting. The most common example of this type is the Bridgeport, described below. Turret mills often have a quill which allows the milling cutter to be raised and lowered in a manner similar to a drill press. This type of machine provides two methods of cutting in the vertical (Z) direction: by raising or lowering the quill, and by moving the knee.

- In the bed mill, however, the table moves only perpendicular to the spindle's axis, while the spindle itself moves parallel to its own axis.

Turret mills are generally considered by some to be more versatile of the two designs. However, turret mills are only practical as long as the machine remains relatively small. As machine size increases, moving the knee up and down requires considerable effort and it also becomes difficult to reach the quill feed handle (if equipped). Therefore, larger milling machines are usually of the bed type.

A third type also exists, a lighter machine, called a mill-drill, which is a close relative of the vertical mill and quite popular with hobbyists. A mill-drill is similar in basic configuration to a small drill press, but equipped with an X-Y table. They also typically use more powerful motors than a comparably sized drill press, with potentiometer-controlled speed and generally have more heavy-duty spindle bearings than a drill press to deal with the lateral loading on the spindle that is created by a milling operation. A mill drill also typically raises and lowers the entire head, including motor, often on a dovetailed vertical, where a drill press motor remains stationary, while the arbor raises and lowers within a driving collar. Other differences that separate a mill-drill from a drill press may be a fine tuning adjustment for the Z-axis, a more precise depth stop, the capability to lock the X, Y or Z axis, and often a system of tilting the head or the entire vertical column and powerhead assembly to allow angled cutting. Aside from size and precision, the principal difference between these hobby-type machines and larger true vertical mills is that the X-Y table is at a fixed elevation; the Z-axis is controlled in basically the same fashion as drill press, where a larger vertical or knee mill has a vertically fixed milling head, and changes the X-Y table elevation. As well, a mill-drill often uses a standard drill press-type Jacob's chuck, rather than an internally tapered arbor that accepts collets. These are frequently of lower quality than other types of machines, but still fill the hobby role well because they tend to be benchtop machines with small footprints and modest price tags.

Horizontal Mill

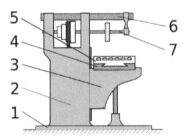

Horizontal milling machine. 1: base 2: column 3: knee 4 & 5: table (x-axis slide is integral) 6: overarm 7: arbor (attached to spindle)

A horizontal mill has the same sort but the cutters are mounted on a horizontal arbor across the table. Many horizontal mills also feature a built-in rotary table that allows milling at various angles; this feature is called a *universal table*. While endmills and the other types of tools available to a vertical mill may be used in a horizontal mill, their real advantage lies in arbor-mounted cutters, called side and face mills, which have a cross section rather like a circular saw, but are generally wider and smaller in diameter. Because the cutters have good support from the arbor and have a larger cross-sectional area than an end mill, quite heavy cuts can be taken enabling rapid material removal rates. These are used to mill grooves and slots. Plain mills are used to shape flat surfaces. Several cutters may be ganged together on the arbor to mill a complex shape of slots and planes. Special cutters can also cut grooves, bevels, radii, or indeed any section desired. These specialty cutters tend to be expensive. Simplex mills have one spindle, and duplex mills have two. It is also easier to cut gears on a horizontal mill. Some horizontal milling machines are equipped with a power-take-off provision on the table. This allows the table feed to be synchronized to a rotary fixture, enabling the milling of spiral features such as hypoid gears.

Comparative Merits

The choice between vertical and horizontal spindle orientation in milling machine design usually hinges on the shape and size of a workpiece and the number of sides of the workpiece that require machining. Work in which the spindle's axial movement is normal to one plane, with an endmill as the cutter, lends itself to a vertical mill, where the operator can stand before the machine and have easy access to the cutting action by looking down upon it. Thus vertical mills are most favored for diesinking work (machining a mould into a block of metal). Heavier and longer workpieces lend themselves to placement on the table of a horizontal mill.

Prior to numerical control, horizontal milling machines evolved first, because they evolved by putting milling tables under lathe-like headstocks. Vertical mills appeared in subsequent decades, and accessories in the form of add-on heads to change horizontal mills to vertical mills (and later vice versa) have been commonly used. Even in the CNC era, a heavy workpiece needing machining on multiple sides lends itself to a horizontal machining center, while diesinking lends itself to a vertical one.

Alternative Classifications

In addition to horizontal versus vertical, other distinctions are also important:

Criterion	Example classification scheme	Comments
Spindle axis orientation	Vertical versus horizontal; Turret versus non-turret	Among vertical mills, "Bridgeport-style" is a whole class of mills inspired by the Bridgeport original, rather like the IBM PC spawned the industry of IBM-compatible PCs by other brands
Control	Manual; Mechanically automated via cams; Digitally automated via NC/CNC	In the CNC era, a very basic distinction is manual versus CNC. Among manual machines, a worthwhile distinction is non-DRO-equipped versus DRO-equipped
Control (specifically among CNC machines)	Number of axes (e.g., 3-axis, 4-axis, or more)	Within this scheme, also: • Pallet-changing versus non-pallet-changing • Full-auto tool-changing versus semi-auto or manual tool-changing
Purpose	General-purpose versus special-purpose or single-purpose	

Criterion	Example classification scheme	Comments
Purpose	Toolroom machine versus production machine	Overlaps with above
Purpose	"Plain" versus "universal"	A distinction whose meaning evolved over decades as technology progressed, and overlaps with other purpose classifications above. Not relevant to today's CNC mills. Regarding manual mills, the common theme is that "plain" mills were production machines with fewer axes than "universal" mills; for example, whereas a plain mill had no indexing head and a non-rotating table, a universal mill would have those. Thus it was suited to universal service, that is, a wider range of possible toolpaths. Machine tool builders no longer use the "plain"-versus-"universal" labeling.
Size	Micro, mini, benchtop, standing on floor, large, very large, gigantic	
Power source	Line-shaft-drive versus individual electric motor drive	Most line-shaft-drive machines, ubiquitous circa 1880–1930, have been scrapped by now
	Hand-crank-power versus electric	Hand-cranked not used in industry but suitable for hobbyist micromills

Variants

A Sieg X2 miniature hobbyist mill plainly showing the basic parts of a mill.

- Bed mill This refers to any milling machine where the spindle is on a *pendant* that moves up and down to move the cutter into the work, while the table sits on a stout *bed* that rests on the floor. These are generally more rigid than a knee mill. Gantry mills can be included in this bed mill category.

- Box mill or column mill Very basic hobbyist bench-mounted milling machines that feature a head riding up and down on a column or box way.

- C-frame mill These are larger, industrial production mills. They feature a knee and fixed spindle head that is only mobile vertically. They are typically much more powerful than a turret mill, featuring a separate hydraulic motor for integral hydraulic power feeds in all directions, and a twenty to fifty horsepower motor. Backlash eliminators are almost always standard equipment. They use large NMTB 40 or 50 tooling. The tables on C-frame mills are usually 18" by 68" or larger, to allow multiple parts to be machined at the same time.

- Floor mill These have a row of rotary tables, and a horizontal pendant spindle mounted on a set of tracks that runs parallel to the table row. These mills have predominantly been converted to CNC, but some can still be found (if one can even find a used machine available) under manual control. The spindle carriage moves to each individual table, performs the machining operations, and moves to the next table while the previous table is being set up for the next operation. Unlike other mills, floor mills have movable floor units. A crane drops massive rotary tables, X-Y tables, etc., into position for machining, allowing large and complex custom milling operations.

- Gantry mill The milling head rides over two rails (often steel shafts) which lie at each side of the work surface.

- Horizontal boring mill Large, accurate bed horizontal mills that incorporate many features from various machine tools. They are predominantly used to create large manufacturing jigs, or to modify large, high precision parts. They have a spindle stroke of several (usually between four and six) feet, and many are equipped with a tailstock to perform very long boring operations without losing accuracy as the bore increases in depth. A typical bed has X and Y travel, and is between three and four feet square with a rotary table or a larger rectangle without a table. The pendant usually provides between four and eight feet of vertical movement. Some mills have a large (30" or more) integral facing head. Right angle rotary tables and vertical milling attachments are available for further flexibility.

- Jig borer Vertical mills that are built to bore holes, and very light slot or face milling. They are typically bed mills with a long spindle throw. The beds are more accurate, and the handwheels are graduated down to .0001" for precise hole placement.

- Knee mill or knee-and-column mill refers to any milling machine whose x-y table rides up and down the column on a vertically adjustable knee. This includes Bridgeports.

- Planer-style mill Large mills built in the same configuration as planers except with a milling spindle instead of a planing head. This term is growing dated as planers themselves are largely a thing of the past.

- Ram-type mill This can refer to any mill that has a cutting head mounted on a sliding ram. The spindle can be oriented either vertically or horizontally. In practice most mills with rams also involve swiveling ability, whether or not it is called "turret" mounting. The Bridgeport configuration can be classified as a vertical-head ram-type mill. Van Norman specialized in ram-type mills through most of the 20th century. Since the wide dissemination of CNC machines, ram-type mills are still made in the Bridgeport configuration (with either manual or CNC control), but the less common variations (such as were built by Van Norman, Index, and others) have died out, their work being done now by either Bridgeport-form mills or machining centers.

- Turret mill More commonly referred to as Bridgeport-type milling machines. The spindle can be aligned in many different positions for a very versatile, if somewhat less rigid machine.

Alternative Terminology

A milling machine is often called a mill by machinists. The archaic term miller was commonly used in the 19th and early 20th centuries.

Since the 1960s there has developed an overlap of usage between the terms milling machine and machining center. NC/CNC machining centers evolved from milling machines, which is why the terminology evolved gradually with considerable overlap that still persists. The distinction, when one is made, is that a machining center is a mill with features that pre-CNC mills never had, especially an automatic tool changer (ATC) that includes a tool magazine (carousel), and sometimes an automatic pallet changer (APC). In typical usage, all machining centers are mills, but not all mills are machining centers; only mills with ATCs are machining centers.

Computer Numerical Control

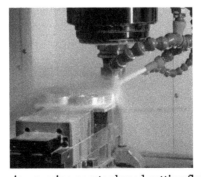

Thin wall milling of aluminum using a water based cutting fluid on the milling cutter

Most CNC milling machines (also called *machining centers*) are computer controlled vertical mills with the ability to move the spindle vertically along the Z-axis. This extra degree of freedom permits their use in diesinking, engraving applications, and 2.5D surfaces such as relief sculptures. When combined with the use of conical tools or a ball nose cutter, it also significantly improves milling precision without impacting speed, providing a cost-efficient alternative to most flat-surface hand-engraving work.

Five-axis machining center with rotating table and computer interface

CNC machines can exist in virtually any of the forms of manual machinery, like horizontal mills. The most advanced CNC milling-machines, the multiaxis machine, add two more axes in addition to the three normal axes (XYZ). Horizontal milling machines also have a C or Q axis, allowing the horizontally mounted workpiece to be rotated, essentially allowing asymmetric and eccentric turning. The fifth axis (B axis) controls the tilt of the tool itself. When all of these axes are used in conjunction with each other, extremely complicated geometries, even organic geometries such as a human head can be made with relative ease with these machines. But the skill to program such geometries is beyond that of most operators. Therefore, 5-axis milling machines are practically always programmed with CAM.

The operating system of such machines is a closed loop system and functions on feedback. These machines have developed from the basic NC (NUMERIC CONTROL) machines. A computerized form of NC machines is known as CNC machines. A set of instructions (called a program) is used to guide the machine for desired operations. Some very commonly used codes, which are used in the program are:

G00 – rapid traverse

G01 – linear interpolation of tool.

G21 – dimensions in metric units.

M03/M04 – spindle start (clockwise/counter clockwise).

T01 M06 – automatic tool change to tool 1

M30 – program end.

Various other codes are also used. A CNC machine is operated by a single operator called a programmer. This machine is capable of performing various operations automatically and economically.

With the declining price of computers and open source CNC software, the entry price of CNC machines has plummeted.

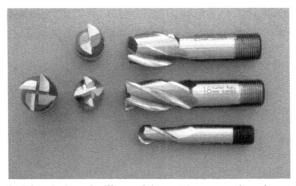

High speed steel with cobalt endmills used for cutting operations in a milling machine.

Tooling

The accessories and cutting tools used on machine tools (including milling machines) are referred to in aggregate by the mass noun "tooling". There is a high degree of standardization of the tooling used with CNC milling machines, and a lesser degree with manual milling machines. To ease up the organization of the tooling in CNC production many companies use a tool management solution.

Milling cutters for specific applications are held in various tooling configurations.

CNC milling machines nearly always use SK (or ISO), CAT, BT or HSK tooling. SK tooling is the most common in Europe, while CAT tooling, sometimes called V-Flange Tooling, is the oldest and probably most common type in the USA. CAT tooling was invented by Caterpillar Inc. of Peoria, Illinois, in order to standardize the tooling used on their machinery. CAT tooling comes in a range of sizes designated as CAT-30, CAT-40, CAT-50, etc. The number refers to the Association for Manufacturing Technology (formerly the National Machine Tool Builders Association (NMTB)) Taper size of the tool.

A CAT-40 toolholder

An improvement on CAT Tooling is BT Tooling, which looks similar and can easily be confused with CAT tooling. Like CAT Tooling, BT Tooling comes in a range of sizes and uses the same NMTB body taper. However, BT tooling is symmetrical about the spindle axis, which CAT tooling is not. This gives BT tooling greater stability and balance at high speeds. One other subtle difference between these two toolholders is the thread used to hold the pull stud. CAT Tooling is all Imperial thread and BT Tooling is all Metric thread. Note that this affects the pull stud only, it does not affect the tool that they can hold, both types of tooling are sold to accept both Imperial and metric sized tools.

A boring head on a Morse taper shank

SK and HSK tooling, sometimes called "Hollow Shank Tooling", is much more common in Europe where it was invented than it is in the United States. It is claimed that HSK tooling is even better than BT Tooling at high speeds. The holding mechanism for HSK tooling is placed within the (hollow) body of the tool and, as spindle speed increases, it expands, gripping the tool more tightly with increasing spindle speed. There is no pull stud with this type of tooling.

For manual milling machines, there is less standardization, because a greater plurality of formerly competing standards exist. Newer and larger manual machines usually use NMTB tooling. This tooling is somewhat similar to CAT tooling but requires a drawbar within the milling machine. Furthermore, there are a number of variations with NMTB tooling that make interchangeability troublesome. The older a machine, the greater the plurality of standards that may apply (e.g., Morse, Jarno, Brown & Sharpe, Van Norman, and other less common builder-specific tapers). However, two standards that have seen especially wide usage are the Morse #2 and the R8, whose prevalence was driven by the popularity of the mills built by Bridgeport Machines of Bridgeport, Connecticut. These mills so dominated the market for such a long time that "Bridgeport" is virtually synonymous with "manual milling machine". Most of the machines that Bridgeport made between 1938 and 1965 used a Morse taper #2, and from about 1965 onward most used an R8 taper.

Accessories

- Arbor support

- Stop block

History

1810s–1830s

Milling machines evolved from the practice of rotary filing—that is, running a circular cutter with file-like teeth in the headstock of a lathe. Rotary filing and, later, true milling were developed to reduce time and effort spent hand-filing. The full story of milling machine development may never be known, because much early development took place in individual shops where few records were kept for posterity. However, the broad outlines are known, as summarized below. From a history-of-technology viewpoint, it is clear that the naming of this new type of machining with the term "milling" was an extension from that word's earlier senses of processing materials by abrading them in some way (cutting, grinding, crushing, etc.).

This milling machine was long credited to Eli Whitney and dated to circa 1818. From the 1910s through the 1940s, this version of its provenance was widely published. In the 1950s and 1960s, various historians of technology mostly discredited the view of this machine as the first miller and possibly even of Whitney as its builder. Nonetheless, it is still an important early milling machine, regardless of its exact provenance.

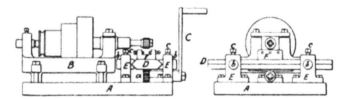

The Middletown milling machine of circa 1818, associated with Robert Johnson and Simeon North.

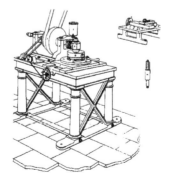

The milling machine built by James Nasmyth between 1829 and 1831 for milling the six sides of a hex nut using an indexing fixture.

Rotary filing long predated milling. A rotary file by Jacques de Vaucanson, circa 1760, is well known. It is clear that milling machines as a distinct class of machine tool (separate from lathes running rotary files) first appeared between 1814 and 1818. The centers of earliest development of true milling machines were two federal armories of the U.S. (Springfield and Harpers Ferry) together with the various private armories and inside contractors that shared turnover of skilled workmen with them.

Between 1912 and 1916, Joseph W. Roe, a respected founding father of machine tool historians, credited Eli Whitney (one of the private arms makers mentioned above) with producing the first true milling machine. By 1918, he considered it "Probably the first milling machine ever built—certainly the oldest now in existence [...]." However, subsequent scholars, including Robert S. Woodbury and others, have improved upon Roe's early version of the history and suggest that just as much credit—in fact, probably more—belongs to various other inventors, including Robert Johnson of Middletown, Connecticut; Captain John H. Hall of the Harpers Ferry armory; Simeon North of the Staddle Hill factory in Middletown; Roswell Lee of the Springfield armory; and Thomas Blanchard. (Several of the men mentioned above are sometimes described on the internet as "the inventor of the first milling machine" or "the inventor of interchangeable parts". Such claims are oversimplified, as these technologies evolved over time among many people.)

Peter Baida, citing Edward A. Battison's article "Eli Whitney and the Milling Machine," which was published in the *Smithsonian Journal of History* in 1966, exemplifies the dispelling of the "Great Man" image of Whitney by historians of technology working in the 1950s and 1960s. He quotes Battison as concluding that "There is no evidence that Whitney developed or used a true milling machine." Baida says, "The so-called Whitney machine of 1818 seems actually to have been made after Whitney's death in 1825." Baida cites Battison's suggestion that the first true milling machine was made not by Whitney, but by Robert Johnson of Middletown.

The late teens of the 19th century were a pivotal time in the history of machine tools, as the period of 1814 to 1818 is also the period during which several contemporary pioneers (Fox, Murray, and Roberts) were developing the planer, and as with the milling machine, the work being done in various shops was undocumented for various reasons (partially because of proprietary secrecy, and also simply because no one was taking down records for posterity).

James Nasmyth built a milling machine very advanced for its time between 1829 and 1831. It was tooled to mill the six sides of a hex nut that was mounted in a six-way indexing fixture.

A milling machine built and used in the shop of Gay & Silver (aka Gay, Silver, & Co) in the 1830s was influential because it employed a better method of vertical positioning than earlier machines. For example, Whitney's machine (the one that Roe considered the very first) and others did not make provision for vertical travel of the knee. Evidently, the workflow assumption behind this was that the machine would be set up with shims, vise, etc. for a certain part design, and successive parts did not require vertical adjustment (or at most would need only shimming). This indicates that early thinking about milling machines was as production machines, not toolroom machines.

In these early years, milling was often viewed as only a roughing operation to be followed by finishing with a hand file. The idea of *reducing* hand filing was more important than *replacing* it.

1840s–1860

Some of the key men in milling machine development during this era included Frederick W. Howe, Francis A. Pratt, Elisha K. Root, and others. (These same men during the same era were also busy developing the state of the art in turret lathes. Howe's experience at Gay & Silver in the 1840s acquainted him with early versions of both machine tools. His machine tool designs were later built at Robbins & Lawrence, the Providence Tool Company, and Brown & Sharpe.) The most successful milling machine design to emerge during this era was the Lincoln miller, which rather than being a specific make and model of machine tool is truly a family of tools built by various companies on a common configuration over several decades. It took its name from the first company to put one on the market, George S. Lincoln & Company (formerly the Phoenix Iron Works), whose first one was built in 1855 for the Colt armory.

A typical Lincoln miller. The configuration was established in the 1850s. (This example was built by Pratt & Whitney, probably 1870s or 1880s.)

During this era there was a continued blind spot in milling machine design, as various designers failed to develop a truly simple and effective means of providing slide travel in all three of the archetypal milling axes (X, Y, and Z—or as they were known in the past, longitudinal, traverse, and vertical). Vertical positioning ideas were either absent or underdeveloped. The Lincoln miller's spindle could be raised and lowered, but the original idea behind its positioning was to be set up in position and then run, as opposed to being moved frequently while running. Like a turret lathe, it was a repetitive-production machine, with each skilled setup followed by extensive fairly low skill operation.

1860s

In 1861, Frederick W. Howe, while working for the Providence Tool Company, asked Joseph R. Brown of Brown & Sharpe for a solution to the problem of milling spirals, such as the flutes of twist drills. These were usually filed by hand at the time. (Helical planing existed but was by no means common.) Brown designed a "universal milling machine" that, starting from its first sale in March 1862, was wildly successful. It solved the problem of 3-axis travel (i.e., the axes that we now call XYZ) much more elegantly than had been done in the past, and it allowed for the milling of spirals using an indexing head fed in coordination with the table feed. The term "universal" was applied to it because it was ready for any kind of work, including toolroom work, and was not as limited in application as previous designs. (Howe had designed a "universal miller" in 1852, but Brown's of 1861 is the one considered a groundbreaking success.)

Brown & Sharpe's groundbreaking universal milling machine, 1861

Brown also developed and patented (1864) the design of formed milling cutters in which successive sharpenings of the teeth do not disturb the geometry of the form.

The advances of the 1860s opened the floodgates and ushered in modern milling practice.

1870s to World War I

In these decades, Brown & Sharpe and the Cincinnati Milling Machine Company dominated the milling machine field. However, hundreds of other firms also built milling machines at the time, and many were significant in various ways. Besides a wide variety of specialized production machines, the archetypal multipurpose milling machine of the late 19th and early 20th centuries was a heavy knee-and-column horizontal-spindle design with power table feeds, indexing head, and a stout overarm to support the arbor. The evolution of machine design was driven not only by inventive spirit but also by the constant evolution of milling cutters that saw milestone after milestone from 1860 through World War I.

A typical universal milling machine of the early 20th century. Suitable for toolroom, jobbing, or production use.

World War I and Interwar Period

Around the end of World War I, machine tool control advanced in various ways that laid the groundwork for later CNC technology. The jig borer popularized the ideas of coordinate dimensioning (dimensioning of all locations on the part from a single reference point); working routinely in "tenths" (ten-thousandths of an inch, 0.0001") as an everyday machine capability; and using the control to go straight from drawing to part, circumventing jig-making. In 1920 the new tracer design of J.C. Shaw was applied to Keller tracer milling machines for die-sinking via the three-dimensional copying of a template. This made diesinking faster and easier just as dies were in higher demand than ever before, and was very helpful for large steel dies such as those used to stamp sheets in automobile manufacturing. Such machines translated the tracer movements to input for servos that worked the machine leadscrews or hydraulics. They also spurred the development of antibacklash leadscrew nuts. All of the above concepts were new in the 1920s but became routine in the NC/CNC era. By the 1930s, incredibly large and advanced milling machines existed, such as the Cincinnati Hydro-Tel, that presaged today's CNC mills in every respect except for CNC control itself.

Bridgeport Milling Machine

In 1936, Rudolph Bannow (1897–1962) conceived of a major improvement to the milling machine. His company commenced manufacturing a new knee-and-column vertical mill in 1938. This was the Bridgeport milling machine, often called a ram-type or turret-type mill because its head has sliding-ram and rotating-turret mounting. The machine became so popular that many other manufacturers created copies and variants. Furthermore, its name came to connote any such variant. The Bridgeport offered enduring advantages over previous models. It was small enough, light enough, and affordable enough to be a practical acquisition for even the smallest machine shop businesses, yet it was also smartly designed, versatile, well-built, and rigid. Its various directions of sliding and pivoting movement allowed the head to approach the work from any angle. The Bridgeport's design became the dominant form for manual milling machines used by several generations of small- and medium-enterprise machinists. By the 1980s an estimated quarter-million Bridgeport milling machines had been built, and they (and their clones) are still being produced today.

1940s–1970s

By 1940, automation via cams, such as in screw machines and automatic chuckers, had already been very well developed for decades. Beginning in the 1930s, ideas involving servomechanisms had been in the air, but it was especially during and immediately after World War II that they began to germinate. These were soon combined with the emerging technology of digital computers. This technological development milieu, spanning from the immediate pre–World War II period into the 1950s, was powered by the military capital expenditures that pursued contemporary advancements in the directing of gun and rocket artillery and in missile guidance—other applications in which humans wished to control the kinematics/dynamics of large machines quickly, precisely, and automatically. Sufficient R&D spending probably would not have happened within the machine tool industry alone; but it was for the latter applications that the will and ability to spend was available. Once the development was underway, it was eagerly applied to machine tool control in one of the many post-WWII instances of technology transfer.

In 1952, numerical control reached the developmental stage of laboratory reality. The first NC machine tool was a Cincinnati Hydrotel milling machine retrofitted with a scratch-built NC control unit. It was reported in *Scientific American,* just as another groundbreaking milling machine, the Brown & Sharpe universal, had been in 1862.

During the 1950s, numerical control moved slowly from the laboratory into commercial service. For its first decade, it had rather limited impact outside of aerospace work. But during the 1960s and 1970s, NC evolved into CNC, data storage and input media evolved, computer processing power and memory capacity steadily increased, and NC and CNC machine tools gradually disseminated from an environment of huge corporations and mainly aerospace work to the level of medium-sized corporations and a wide variety of products. NC and CNC's drastic advancement of machine tool control deeply transformed the culture of manufacturing.

1980s–present

Computers and CNC machine tools continue to develop rapidly. The personal computer revolution has a great impact on this development. By the late 1980s small machine shops had desktop computers and CNC machine tools. Soon after, hobbyists, artists, and designers began obtaining CNC mills and lathes. Manufacturers have started producing economically priced CNCs machines small enough to sit on a desktop which can cut at high resolution materials softer than stainless steel. They can be used to make anything from jewelry to printed circuit boards to gun parts, even fine art.

Milling Standards

National and international standards are used to standardize the definitions, environmental requirements, and test methods used for milling. Selection of the standard to be used is an agreement between the supplier and the user and has some significance in the design of the mill. In the United States, ASME has developed the B5.45-1972 Standard entitled "Milling Machines", B94.19-1997, Milling Cutters and End Mills.

Grinding (Abrasive Cutting)

Grinding is an abrasive machining process that uses a grinding wheel as the cutting tool.

Grinding

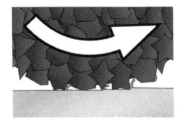

Sketch of how abrasive particles in a grinding wheel remove material from a workpiece.

A wide variety of machines are used for grinding:

- Hand-cranked knife-sharpening stones (grindstones)
- Handheld power tools such as angle grinders and die grinders

- Various kinds of expensive industrial machine tools called grinding machines

- Bench grinders often found in residential garages and basements

Grinding practice is a large and diverse area of manufacturing and toolmaking. It can produce very fine finishes and very accurate dimensions; yet in mass production contexts it can also rough out large volumes of metal quite rapidly. It is usually better suited to the machining of very hard materials than is "regular" machining (that is, cutting larger chips with cutting tools such as tool bits or milling cutters), and until recent decades it was the only practical way to machine such materials as hardened steels. Compared to "regular" machining, it is usually better suited to taking very shallow cuts, such as reducing a shaft's diameter by half a thousandth of an inch or 12.7 µm.

Grinding is a subset of cutting, as grinding is a true metal-cutting process. Each grain of abrasive functions as a microscopic single-point cutting edge (although of high negative rake angle), and shears a tiny chip that is analogous to what would conventionally be called a "cut" chip (turning, milling, drilling, tapping, etc.). However, among people who work in the machining fields, the term *cutting* is often understood to refer to the macroscopic cutting operations, and *grinding* is often mentally categorized as a "separate" process. This is why the terms are usually used in separately in shop-floor practice.

Lapping and sanding are subsets of grinding.

Processes

Selecting which of the following grinding operations to be used is determined by the size, shape, features and the desired production rate.

Surface Grinding

Surface grinding uses a rotating abrasive wheel to remove material, creating a flat surface. The tolerances that are normally achieved with grinding are $\pm 2 \times 10^{-4}$ inches for grinding a flat material, and $\pm 3 \times 10^{-4}$ inches for a parallel surface (in metric units: 5 µm for flat material and 8 µm for parallel surface).

The surface grinder is composed of an abrasive wheel, a workholding device known as a chuck, either electromagnetic or vacuum, and a reciprocating table.

Grinding is commonly used on cast iron and various types of steel. These materials lend themselves to grinding because they can be held by the magnetic chuck commonly used on grinding machines, and they do not melt into the wheel, clogging it and preventing it from cutting. Materials that are less commonly ground are Aluminum, stainless steel, brass & plastics. These all tend to clog the cutting wheel more than steel & cast iron, but with special techniques it is possible to grind them.

Cylindrical Grinding

Cylindrical grinding (also called center-type grinding) is used to grind the cylindrical surfaces and shoulders of the workpiece. The workpiece is mounted on centers and rotated by a device known as a drive dog or center driver. The abrasive wheel and the workpiece are rotated by separate motors and at different speeds. The table can be adjusted to produce tapers. The wheel head can be swiveled. The five types of cylindrical grinding are: outside diameter (OD) grinding, inside diameter (ID) grinding, plunge grinding, creep feed grinding, and centerless grinding.

A cylindrical grinder has a grinding (abrasive) wheel, two centers that hold the workpiece, and a chuck, grinding dog, or other mechanism to drive the work. Most cylindrical grinding machines include a swivel to allow for the forming of tapered pieces. The wheel and workpiece move parallel to one another in both the radial and longitudinal directions. The abrasive wheel can have many shapes. Standard disk-shaped wheels can be used to create a tapered or straight workpiece geometry while formed wheels are used to create a shaped workpiece. The process using a formed wheel creates less vibration than using a regular disk-shaped wheel.

Tolerances for cylindrical grinding are held within five ten-thousandths of an inch (± 0.0005) (metric: ± 13 um) for diameter and one ten-thousandth of an inch(± 0.0001) (metric: 2.5 um) for roundness. Precision work can reach tolerances as high as fifty millionths of an inch (± 0.00005) (metric: 1.3 um) for diameter and ten millionths (± 0.00001) (metric: 0.25 um) for roundness. Surface finishes can range from 2 to 125 microinches (metric: 50 nm to 3 um), with typical finishes ranging from 8 to 32 microinches. (metric: 0.2 um to 0.8 um)

Creep-feed Grinding

Creep-feed grinding (CFG) was invented in Germany in the late 1950s by Edmund and Gerhard Lang. Unlike normal grinding, which is used primarily to finish surfaces, CFG is used for high rates of material removal, competing with milling and turning as a manufacturing process choice. Depths of cut of up to 6 mm (0.25 inches) are used along with low workpiece speed. Surfaces with a softer-grade resin bond are used to keep workpiece temperature low and an improved surface finish up to 1.6 micrometres Rmax

With CFG it takes 117 sec to remove 1 in.^{3} of material, whereas precision grinding would take more than 200 sec to do the same. CFG has the disadvantage of a wheel that is constantly degrading, and requires high spindle power, 51 hp (38 kW), and is limited in the length of part it can machine.

To address the problem of wheel sharpness, continuous-dress creep-feed grinding (CDCF) was developed in the 1970s. It dresses the wheel constantly during machining, keeping it in a state of specified sharpness. It takes only 17 sec. to remove 1 in^3 of material, a huge gain in productivity. 38 hp (28 kW) spindle power is required, and runs at low to conventional spindle speeds. The limit on part length was erased.

High-efficiency deep grinding (HEDG) uses plated superabrasive wheels, which never need dressing and last longer than other wheels. This reduces capital equipment investment costs. HEDG can be used on long part lengths, and removes material at a rate of 1 in³ in 83 sec. It requires high spindle power and high spindle speeds.

Peel grinding, patented under the name of Quickpoint in 1985 by Erwin Junker Maschinenfabrik, GmbH in Nordrach, Germany, uses a tool with a superabrasive nose and can machine cylindrical parts.

Ultra-high speed grinding (UHSG) can run at speeds higher than 40,000 fpm (200 m/s), taking 41 sec to remove 1 in.³ of material, but is still in the R&D stage. It also requires high spindle power and high spindle speeds.

Others

Form grinding is a specialized type of cylindrical grinding where the grinding wheel has the exact shape of the final product. The grinding wheel does not traverse the workpiece.

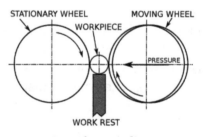

Centerless grinding

Internal grinding is used to grind the internal diameter of the workpiece. Tapered holes can be ground with the use of internal grinders that can swivel on the horizontal.

Centerless grinding is when the workpiece is supported by a blade instead of by centers or chucks. Two wheels are used. The larger one is used to grind the surface of the workpiece and the smaller wheel is used to regulate the axial movement of the workpiece. Types of centerless grinding include through-feed grinding, in-feed/plunge grinding, and internal centerless grinding.

Pre-grinding When a new tool has been built and has been heat-treated, it is pre-ground before welding or hardfacing commences. This usually involves grinding the OD slightly higher than the finish grind OD to ensure the correct finish size.

Electrochemical grinding is a type of grinding in which a positively charged workpiece in a conductive fluid is eroded by a negatively charged grinding wheel. The pieces from the workpiece are dissolved into the conductive fluid.

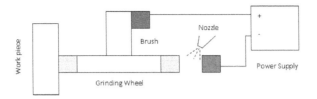

A schematic of ELID grinding

Electrolytic in-process dressing (ELID) grinding is one of the most accurate grinding methods. In this ultra precision grinding technology the grinding wheel is dressed electrochemically and in-process to maintain the accuracy of the grinding. An ELID cell consists of a metal bonded grinding wheel, a cathode electrode, a pulsed DC power supply and electrolyte. The wheel is connected to the positive terminal of the DC power supply through a carbon brush whereas the electrode is connected to the negative pole of the power supply. Usually alkaline liquids are used as both electrolytes and coolant for grinding. A nozzle is used to inject the electrolyte into the gap between wheel and electrode. The gap is usually maintained to be approximately 0.1mm to 0.3 mm. During the grinding operation one side of the wheel takes part in the grinding operation whereas the other side of the wheel is being dressed by electrochemical reaction. The dissolution of the metallic bond material is caused by the dressing which in turns results continuous protrusion of new sharp grits.

Grinding Wheel

A grinding wheel is an expendable wheel used for various grinding and abrasive machining operations. It is generally made from a matrix of coarse abrasive particles pressed and bonded together to form a solid, circular shape, various profiles and cross sections are available depending on the intended usage for the wheel. Grinding wheels may also be made from a solid steel or aluminium disc with particles bonded to the surface.

Lubrication

The use of fluids in a grinding process is necessary to cool and lubricate the wheel and workpiece as well as remove the chips produced in the grinding process. The most common grinding fluids are water-soluble chemical fluids, water-soluble oils, synthetic oils, and petroleum-based oils. It is imperative that the fluid be applied directly to the cutting area to prevent the fluid being blown away from the piece due to rapid rotation of the wheel.

Work Material	Cutting Fluid	Application
Aluminum	Light-duty oil	Flood
Brass	Light-duty oil	Flood

Cast Iron	Heavy-duty emulsifiable oil, light-duty chemical oil, synthetic oil	Flood
Mild Steel	Heavy-duty water-soluble oil	Flood
Stainless Steel	Heavy-duty emulsifiable oil, heavy-duty chemical oil, synthetic oil	Flood
Plastics	Water-soluble oil, heavy-duty emulsifiable oil, dry, light-duty chemical oil, synthetic oil	Flood

The Workpiece

Workholding Methods

The workpiece is manually clamped to a lathe dog, powered by the faceplate, that holds the piece in between two centers and rotates the piece. The piece and the grinding wheel rotate in opposite directions and small bits of the piece are removed as it passes along the grinding wheel. In some instances special drive centers may be used to allow the edges to be ground. The workholding method affects the production time as it changes set up times.

Workpiece Materials

Typical workpiece materials include aluminum, brass, plastics, cast iron, mild steel, and stainless steel. Aluminum, brass and plastics can have poor to fair machinability characteristics for cylindrical grinding. Cast Iron and mild steel have very good characteristics for cylindrical grinding. Stainless steel is very difficult to grind due to its toughness and ability to work harden, but can be worked with the right grade of grinding wheels.

Workpiece Geometry

The final shape of a workpiece is the mirror image of the grinding wheel, with cylindrical wheels creating cylindrical pieces and formed wheels creating formed pieces. Typical sizes on workpieces range from .75 in. to 20 in. (metric: 18mm to 1 m) and .80 in. to 75 in. in length (metric: 2 cm to 4 m), although pieces between .25 in. and 60 in. in diameter (metric: 6 mm to 1.5 m) and .30 in. and 100 in. in length (metric: 8 mm to 2.5 m) can be ground. Resulting shapes can range from straight cylinders, straight edged conical shapes, or even crankshafts for engines that experience relatively low torque.

Effects on Workpiece Materials

Mechanical properties will change due to stresses put on the part during finishing. High grinding temperatures may cause a thin martensitic layer to form on the part, which will lead to reduced material strength from microcracks.

Physical property changes include the possible loss of magnetic properties on ferromagnetic materials.

Chemical property changes include an increased susceptibility to corrosion because of high surface stress.

Abrasive Machining

Abrasive machining is a machining process where material is removed from a workpiece using a multitude of small abrasive particles. Common examples include grinding, honing, and polishing. Abrasive processes are usually expensive, but capable of tighter tolerances and better surface finish than other machining processes

Mechanics of Abrasive Machining

Abrasive machining works by forcing the abrasive particles, or grains, into the surface of the workpiece so that each particle cuts away a small bit of material. Abrasive machining is similar to conventional machining, such as milling or turning, because each of the abrasive particles acts like a miniature cutting tool. However, unlike conventional machining the grains are much smaller than a cutting tool, and the geometry and orientation of individual grains are not well defined. As a result, abrasive machining is less power efficient and generates more heat. The grain size may be different based on the machining. For rough grinding, coarse abrasives are used. For fine grinding, fine grains (abrasives) are used.

Abrasive Machining Processes

Abrasive machining processes can be divided into two categories based on how the grains are applied to the workpiece.

In bonded abrasive processes, the particles are held together within a matrix, and their combined shape determines the geometry of the finished workpiece. For example, in grinding the particles are bonded together in a wheel. As the grinding wheel is fed into the part, its shape is transferred onto the workpiece.

In loose abrasive processes, there is no structure connecting the grains. They may be applied without lubrication as dry powder, or they may be mixed with a lubricant to form a slurry. Since the grains can move independently, they must be forced into the workpiece with another object like a polishing cloth or a lapping plate.

Common abrasive processes are listed below.

Fixed (Bonded) Abrasive Processes

- Grinding
- Honing, superfinishing

- Tape finishing, abrasive belt machining

- Abrasive sawing, Diamond wire cutting, Wire saw

- Sanding

Loose Abrasive Processes

- Polishing

- Lapping

- Abrasive flow machining (AFM)

- Hydro-erosive grinding

- Water-jet cutting

- Abrasive blasting

- Mass finishing, tumbling

    o   Open barrel tumbling

    o   Vibratory bowl tumbling

    o   Centrifugal disc tumbling

    o   Centrifugal barrel tumbling

Abrasives

The most important property of an abrasive is its hardness. For abrasive grains to effectively cut, they must be significantly harder than the workpiece material. They can be grouped based on their hardness into two categories: conventional abrasives and superabrasives.

Conventional abrasive materials have been used by man since the advent of machining. They are made of materials that exist naturally on Earth, and they are abundant and cheap. Conventional abrasives can suitably machine most materials.

Superabrasives are much harder than conventional abrasives. Since they are much more expensive, they are used when conventional abrasives will not suffice.

Common abrasives are listed below.

Conventional

- Aluminum oxide (Corundum)

- Silicon carbide

- Emery

- Pumice

- Sand

- Steel abrasive

Superabrasives

- Diamond

- Cubic Boron Nitride (CBN), Borazon

Abrasive Jet Machining

Abrasive jet machining (AJM), also known as abrasive micro-blasting, pencil blasting and micro-abrasive blasting, is an abrasive blasting machining process that uses abrasives propelled by a high velocity gas to erode material from the workpiece. Common uses include cutting heat-sensitive, brittle, thin, or hard materials. Specifically it is used to cut intricate shapes or form specific edge shapes.

Process

Material is removed by fine abrasive particles, usually about 0.001 in (0.025 mm) in diameter, driven by a high velocity fluid stream; common gases are air or inert gases. Pressures for the gas range from 25 to 130 psig (170–900 kPag) and speeds can be as high as 300 m/s (1,000 kph).

Equipment

AJM machines are usually self-contained bench-top units. First it compresses the gas and then mixes it with the abrasive in a mixing chamber. The gas passes through a convergent-divergent nozzle before entering the mixing chamber, and then exits through a convergent nozzle. The nozzle can be hand held or mounted in a fixture for automatic operations.

Nozzles must be highly resistant to abrasion and are typically made of tungsten carbide or synthetic sapphire. For average material removal, tungsten carbide nozzles have a useful life of 12 to 30 hours, and sapphire nozzles last about 300 hours. The distance of the nozzle from the workpiece affects the size of the machined area and the rate of material removal.

Grit size and orifice diameters for various abrasive materials		
Abrasive material	Grit size (μin)	Orifice diameter (in)
Aluminum oxide	10–50	0.005–0.018
Silicon carbide	25–50	0.008–0.018
Glass beads	2500	0.026–0.05

Advantages and Disadvantages

The main advantages are its flexibility, low heat production, and ability to machine hard and brittle materials. Its flexibility owes from its ability to use hoses to transport the gas and abrasive to any part of the workpiece.

One of the main disadvantages is its slow material removal rate; for this reason it is usually used as a finishing process. Another disadvantage is that the process produces a tapered cut.

Productivity Improving Technologies

Productivity in general is a ratio of output to input in the production of goods and services. Productivity is increased by lowering the amount of labor, capital, energy or materials that go into producing economic goods. Increases in productivity are largely responsible for the increase in per capita living standards.

The Spinning Jenny and Spinning Mule (shown) greatly increased the productivity of thread manufacturing compared to the spinning wheel.

History

Productivity improving technologies date back to antiquity, with rather slow progress until the late Middle Ages. Important examples of early to European medieval technology include the water wheel, the horse collar, the spinning wheel, the three field system (after 1500 the four field system) and the blast furnace. All of these technologies had been in use in China, some for centuries, before being introduced to Europe.

Technological progress was aided by literacy and the diffusion of knowledge that accelerated after the spinning wheel spread to Western Europe in the 13th century. The spinning wheel increased the supply of rags used for pulp in paper making, whose technology reached Sicily sometime in the 12th century. Cheap paper was a factor in the development of the movable type printing press, which lead to a large increase in the number of books and titles published. Books on science and technology eventually began to appear, such as the mining technical manual *De Re Metallica*, which was the most important technology book of the 16th century and was the standard chemistry text for the next 180 years.

Francis Bacon (1561-1626) is known for the scientific method, which was a key factor in the scientific revolution. Bacon stated that the technologies that distinguished Europe of his day from the Middle Ages were paper and printing, gunpowder and the magnetic compass, known as the Four great inventions. The Four great inventions important to the development of Europe were of Chinese origin. Other Chinese inventions included the horse collar, cast iron, an improved plow and the seed drill.

Mining and metal refining technologies played a key role in technological progress. Much of our understanding of fundamental chemistry evolved from ore smelting and refining, with *De Re Metallica* being the leading chemistry text for 180 years. Railroads evolved from mine carts and the first steam engines were designed specifically for pumping water from mines. The significance of the blast furnace goes far beyond its capacity for large scale production of cast iron. The blast furnace was the first example of continuous production and is a countercurrent exchange process, various types of which are also used today in chemical and petroleum refining. Hot blast, which recycled what would have otherwise been waste heat, was one of engineering's key technologies. It had the immediate effect of dramatically reducing the energy required to produce pig iron, but reuse of heat was eventually applied to a variety of industries, particularly steam boilers, chemicals, petroleum refining and pulp and paper.

Before the 17th century scientific knowledge tended to stay within the intellectual community, but by this time it became accessible to the public in what is called "open science". Near the beginning of the Industrial Revolution came publication of the Encyclopédie, written by numerous contributors and edited by Denis Diderot and Jean le Rond d'Alembert (1751–72). It contained many articles on science and was the first general encyclopedia to provide in depth coverage on the mechanical arts, but is far more recognized for its presentation of thoughts of the Enlightenment.

Economic historians generally agree that, with certain exceptions such as the steam engine, there is no strong linkage between the 17th century scientific revolution (Descartes, Newton, etc.) and the Industrial Revolution. However, an important mechanism for the transfer of technical knowledge was scientific societies, such as The Royal Society of London for Improving Natural Knowledge, better known as the Royal Society, and the Académie des Sciences. There were also technical colleges, such as the École Polytechnique. Scotland was the first place where science was taught (in the 18th century) and was where Joseph Black discovered heat capacity and latent heat and where his friend James Watt used knowledge of heat to conceive the separate condenser as a means to improve the efficiency of the steam engine.

Probably the first period in history in which economic progress was observable after one generation was during the British Agricultural Revolution in the 18th century. However, technological and economic progress did not proceed at a significant rate until the English Industrial Revolution in the late 18th century, and even then productivity grew about 0.5% annually. High productivity growth began during the late 19th century in what is sometimes call the Second Industrial Revolution. Most major innovations of the Second Industrial Revolution were based on the modern scientific understanding of chemistry, electromagnetic theory and thermodynamics and other principles known to a new profession of engineering.

Major Sources of Productivity Growth in Economic History

1900s photograph of barge pullers on the Volga River. Pushing was done with poles and manual pulling using overhanging tree branches. Horses were also used.

New Forms of Energy and Power

Before the industrial revolution the only sources of power were water, wind and muscle. Most good water power sites (those not requiring massive modern dams) in Europe were developed during the medieval period. In the 1750s John Smeaton, the "father of civil engineering," significantly improved the efficiency of the water wheel by applying scientific principles, thereby adding badly needed power for the Industrial Revolution. However water wheels remained costly, relatively inefficient and not well suited to very large power dams. Benoît Fourneyron's highly efficient turbine developed in the late 1820s eventually replaced waterwheels. Fourneyron type turbines can operate at 95% efficiency and used in today's large hydro-power installations. Hydro-power continued to be the leading source of industrial power in the United States until past the mid 19th century because of abundant sites, but steam power overtook water power in the UK decades earlier.

In 1711 a Newcomen steam engine was installed for pumping water from a mine, a job that typically was done by large teams of horses, of which some mines used as many as 500. Animals convert feed to work at an efficiency of about 5%, but while this was much more than the less than 1% efficiency of the early Newcomen engine, in coal mines there was low quality coal with little market value available. Fossil fuel energy first exceeded all animal and water power in 1870. The role energy and machines replacing physical work is discussed in Ayres-Warr (2004, 2009).

While steamboats were used in some areas, as recently as the late 19th Century thousands of workers pulled barges. Until the late 19th century most coal and other minerals were mined with picks and shovels and crops were harvested and grain threshed using animal power or by hand. Heavy loads like 382 pound bales of cotton were handled on hand trucks until the early 20th century.

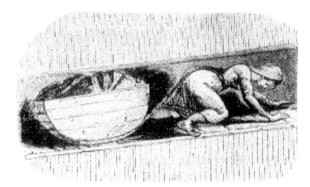

A young "drawer" pulling a coal tub along a mine gallery. Minecarts were more common than the skid shown. Railroads descended from minecarts. In Britain laws passed in 1842 and 1844 improved working conditions in mines.

Excavation was done with shovels until the late 19th century when steam shovels came into use. It was reported that a laborer on the western division of the Erie Canal was expected to dig 5 cubic yards per day in 1860; however, by 1890 only 3-1/2 yards per day were expected. Today's large electric shovels have buckets that can hold 168 cubic meters and consume the power of a city of 100,000.

Dynamite, a safe to handle blend of nitroglycerin and diatomaceous earth was patented in 1867 by Alfred Nobel. Dynamite increased productivity of mining, tunneling, road building, construction and demolition and made projects such as the Panama Canal possible.

Steam power was applied to threshing machines in the late 19th century. There were steam engines that moved around on wheels under their own power that were used for supplying temporary power to stationary farm equipment such as threshing machines. These were called *road engines,* and Henry Ford seeing one as a boy was inspired to build an automobile. Steam tractors were used but never became popular.

With internal combustion came the first mass-produced tractors (Fordson c. 1917). Tractors replaced horses and mules for pulling reapers and combine harvesters, but in the 1930s self powered combines were developed. Output per man hour in growing wheat rose by a factor of about 10 from the end of World War II until about 1985, largely because of powered machinery, but also because of increased crop yields. Corn manpower showed a similar but higher productivity increase.

One of the greatest periods of productivity growth coincided with the electrification of factories which took place between 1900 and 1930 in the U.S.

Energy Efficiency

In engineering and economic history the most important types of energy efficiency were in the conversion of heat to work, the reuse of heat and the reduction of friction. There was also a dramatic reduction energy required to transmit electronic signals, both voice and data.

Conversion of Heat to Work

The early Newcomen steam engine was about 0.5% efficient and was improved to slightly over 1% by John Smeaton before Watt's improvements, which increased thermal efficiency to 2%. In 1900 it took 7 lbs coal/ kw hr.

Electrical generation was the sector with the highest productivity growth in the U.S. in the early twentieth century. After the turn of the century large central stations with high pressure boilers and efficient steam turbines replaced reciprocating steam engines and by 1960 it took 0.9 lb coal per kw-hr. Counting the improvements in mining and transportation the total improvement was by a factor greater than 10. Today's steam turbines have efficiencies in the 40% range. Most electricity today is produced by thermal power stations using steam turbines.

The Newcomen and Watt engines operated near atmospheric pressure and used atmospheric pressure, in the form of a vacuum caused by condensing steam, to do work. Higher pressure engines were light enough, and efficient enough to be used for powering ships and locomotives. Multiple expansion (multi-stage) engines were developed in the 1870s and were efficient enough for the first time to allow ships to carry more freight than coal, leading to great increases in international trade.

The first important diesel ship was the *MS Selandia* launched in 1912. By 1950 one-third of merchant shipping was diesel powered. Today the most efficient prime mover is the two stroke marine diesel engine developed in the 1920s, now ranging in size to over 100,000 horsepower with a thermal efficiency of 50%.

Steam locomotives that used up to 20% of the U.S. coal production were replaced by diesel locomotives after World War II, saving a great deal of energy and reducing manpower for handling coal, boiler water and mechanical maintenance.

Improvements in steam engine efficiency caused a large increase in the number of steam engines and the amount of coal used, as noted by William Stanley Jevons in *The Coal Question*. This is called the Jevons paradox.

Electrification and The Pre-Electric Transmission of Power

Electricity consumption and economic growth are strongly correlated. Per capita electric consumption correlates almost perfectly with economic development. Electrification was the first technology to enable long distance transmission of power with minimal power losses. Electric motors did away with line shafts for distributing power and dramatically increased the productivity of factories. Very large central power stations created economies of scale and were much more efficient at producing power than reciprocating steam engines. Electric motors greatly reduced the capital cost of power compared to steam engines.

The main forms of pre-electric power transmission were line shafts, hydraulic power networks and pneumatic and wire rope systems. Line shafts were the common form of power transmission in factories from the earliest industrial steam engines until factory electrification. Line shafts limited factory arrangement and suffered from high power losses. Hydraulic power came into use in the mid 19th century. It was used extensively in the Bessemer process and for cranes at ports, especially in the UK. London and a few other cities had hydraulic utilities that provided pressurized water for industrial over a wide area.

Pneumatic power began being used industry and in mining and tunneling in the last quarter of the 19th century. Common applications included rock drills and jack hammers. Wire ropes supported by large grooved wheels were able to transmit power with low loss for a distance of a few miles or kilometers. Wire rope systems appeared shortly before electrification.

Reuse of Heat

Recovery of heat for industrial processes was first widely used as hot blast in blast furnaces to make pig iron in 1828. Later heat reuse included the Siemens-Martin process which was first used for making glass and later for steel with the open hearth furnace. Today heat is reused in many basic industries such as chemicals, oil refining and pulp and paper, using a variety of methods such as heat exchangers in many processes.

Multiple-effect evaporators use vapor from a high temperature effect to evaporate a lower temperature boiling fluid. In the recovery of kraft pulping chemicals the spent black liquor can be evaporated five or six times by reusing the vapor from one effect to boil the liquor in the preceding effect. Cogeneration is a process that uses high pressure steam to generate electricity and then uses the resulting low pressure steam for process or building heat.

Industrial process have undergone numerous minor improvements which collectively made significant reductions in energy consumption per unit of production.

Reducing Friction

Reducing friction was one of the major reasons for the success of railroads compared to wagons. This was demonstrated on an iron plate covered wooden tramway in 1805 at Croydon, U.K.

" A good horse on an ordinary turnpike road can draw two thousand pounds, or one ton. A party of gentlemen were invited to witness the experiment, that the superiority of the new road might be established by ocular demonstration. Twelve wagons were loaded with stones, till each wagon weighed three tons, and the wagons were fastened together. A horse was then attached, which drew the wagons with ease, six miles in two hours, having stopped four times, in order to show he had the power of starting, as well as drawing his great load."

Better lubrication, such as from petroleum oils, reduced friction losses in mills and factories. Anti-friction bearings were developed using alloy steels and precision machining techniques available in the last quarter of the 19th century. Anti-friction bearings were widely used on bicycles by the 1880s. Bearings began being used on line shafts in the decades before factory electrification and it was the pre-bearing shafts that were largely responsible for their high power losses, which were commonly 25 to 30% and often as much as 50%.

Lighting Efficiency

Electric lights were far more efficient than oil or gas lighting and did not generate smoke, fumes nor as much heat. Electric light extended the work day, making factories, businesses and homes more productive. Electric light was not a great fire hazard like oil and gas light.

The efficiency of electric lights has continuously improved from the first incandescent lamps to tungsten filament lights. The fluorescent lamp, which became commercial in the late 1930s, is much more efficient than incandescent lighting. Light-emitting diodes or LED's are highly efficient and long lasting.

Infrastructures

The relative energy required for transport of a tonne-km for various modes of transport are: pipelines=1(basis), water 2, rail 3, road 10, air 100.

Roads

Unimproved roads were extremely slow, costly for transport and dangerous. In the 18th century layered gravel began being increasingly used, with the three layer Macadam coming into use in the early 19th century. These roads were crowned to shed water and had drainage ditches along the sides. The top layer of stones eventually crushed to fines and smoothed the surface somewhat. The lower layers were of small stones that allowed good drainage. Importantly, they offered less resistance to wagon wheels and horses hooves and feet did not sink in the mud. Plank roads also came into use in the U.S. in the 1810s-1820s. Improved roads were costly, and although they cut the cost of land transportation in half or more, they were soon overtaken by railroads as the major transportation infrastructure.

Ocean Shipping and Inland Waterways

Sailing ships could transport goods for over a 3000 miles for the cost of 30 miles by wagon. A horse that could pull a one-ton wagon could pull a 30-ton barge. During the English or First Industrial Revolution, supplying coal to the furnaces at Manchester was difficult because there were few roads and because of the high cost of using wagons. However, canal barges were known to be workable, and this was demonstrated by building the Bridgewater Canal, which opened in 1761, bringing coal from Worsley to Manchester. The Bridgewater Canal's success started a frenzy of canal building that lasted until the appearance of railroads in the 1830s.

Railroads

Railroads greatly reduced the cost of overland transportation. It is estimated that by 1890 the cost of wagon freight was U.S. 24.5 cents/ton-mile versus 0.875 cents/ton-mile by railroad, for a decline of 96%.

Electric street railways (trams, trolleys or streetcars) were in the final phase of railroad building from the late 1890s and first two decades of the 20th century. Street railways were soon displaced by motor buses and automobiles after 1920.

Motorways

Highways with internal combustion powered vehicles completed the mechanization of overland transportation. When trucks appeared c. 1920 the price transporting farm goods to market or to rail stations was greatly reduced. Motorized highway transport also reduced inventories.

The high productivity growth in the U.S. during the 1930s was in large part due to the highway building program of that decade.

Pipelines

Pipelines are the most energy efficient means of transportation. Iron and steel pipelines came into use during latter part of the 19th century, but only became a major infrastructure during the 20th century. Centrifugal pumps and centrifugal compressors are efficient means of pumping liquids and natural gas.

Mechanization

Adriance-Getreidemäher.

Adriance reaper, late 19th century

Threshing machine from 1881. Steam engines were also used instead of horses. Today both threshing and reaping are done with a combine harvester.

Mechanized Agriculture

The seed drill is a mechanical device for spacing and planting seed at the appropriate depth. It originated in ancient China before the 1st century BC. Saving seed was extremely important at a time when yields were measured in terms of seeds harvested per seed planted, which was typically between 3 and 5. The seed drill also saved planting labor. Most importantly, the seed drill meant crops were grown in rows, which reduced competition of plants and increase yields. It was reinvented in 16th century Europe based on verbal descriptions and crude drawings brought back from China. Jethro Tull patented a version in 1700; however, it was expensive and unreliable. Reliable seed drills appeared in the mid 19th century.

Since the beginning of agriculture threshing was done by hand with a flail, requiring a great deal of labor. The threshing machine (ca. 1794) simplified the operation and allowed it to use animal power. By the 1860s threshing machines were widely introduced and ultimately displaced as much as a quarter of agricultural labor. In Europe, many of the displaced workers were driven to the brink of starvation.

Harvesting oats in a Claas Lexion 570 combine with enclosed, air-conditioned cab with rotary thresher and laser-guided hydraulic steering

Before c. 1790 a worker could harvest 1/4 acre per day with a scythe. In the early 1800s the grain cradle was introduced, significantly increasing the productivity of hand labor. It was estimated that each of Cyrus McCormick's horse pulled reapers (Ptd. 1834) freed up five men for military service in the U.S. Civil War. By 1890 two men and two horses could cut, rake and bind 20 acres of wheat per day. In the 1880s the reaper and threshing machine were combined into the combine harvester. These machines required large teams of horses or mules to pull. Over the entire 19th century the output per man hour for producing wheat rose by about 500% and for corn about 250%.

Farm machinery and higher crop yields reduced the labor to produce 100 bushels of corn from 35 to 40 hours in 1900 to 2 hours 45 minutes in 1999. The conversion of agricultural mechanization to internal combustion power began after 1915. The horse population began to decline in the 1920s after the conversion of agriculture and transportation to internal combustion. In addition to saving labor, this freed up much land previously used for supporting draft animals.

The peak years for tractor sales in the U.S. were the 1950s. There was a large surge in horsepower of farm machinery in the 1950s.

Industrial Machinery

The most important mechanical devices before the Industrial Revolution were water and wind mills. Water wheels date to Roman times and windmills somewhat later. Water and wind power were first used for grinding grain into flour, but were later adapted to power trip hammers for pounding rags into pulp for making paper and for crushing ore. Just before the Industrial revolution water power was applied to bellows for iron smelting in Europe. (Water powered blast bellows were used in ancient China.) Wind and water power were also used in sawmills. The technology of building mills and mechanical clocks was important to the development of the machines of the Industrial Revolution.

The spinning wheel was a medieval invention that increased thread making productivity by a factor greater than ten. One of the early developments that preceded the Industrial Revolution was the stocking frame (loom) of c. 1589. Later in the Industrial Revolution came the flying shuttle, a simple device that doubled the productivity of weaving. Spinning thread had been a limiting factor in cloth making requiring 10 spinners using the spinning wheel to supply one weaver. With the spinning jenny a spinner could spin eight threads at once. The water frame (Ptd. 1768) adapted water power to spinning, but it could only spin one thread at a time. The water frame was easy to operate and many could be located in a single building. The spinning mule (1779) allowed a large number of threads to be spun by a single machine using water power. A change in consumer preference for cotton at the time of increased cloth production resulted in the invention of the cotton gin (Ptd. 1794). Steam power eventually was used as a supplement to water during the Industrial Revolution, and both were used until electrification. A graph of productivity of spinning technologies can be found in Ayres (1989), along with much other data related this article.

With a cotton gin (1792) in one day a man could remove seed from as much upland cotton as would have previously taken a woman working two months to process at one pound per day using a roller gin.

An early example of a large productivity increase by special purpose machines is the c. 1803 Portsmouth Block Mills. With these machines 10 men could produce as many blocks as 110 skilled craftsmen.

In the 1830s several technologies came together to allow an important shift in wooden building construction. The circular saw (1777), cut nail machines (1794), and steam engine allowed slender pieces of lumber such as 2"x4"s to be efficiently produced and then nailed together in what became known as balloon framing (1832). This was the beginning of the decline of the ancient method of timber frame construction with wooden joinery.

Following mechanization in the textile industry was mechanization of the shoe industry.

The sewing machine, invented and improved during the early 19th century and produced in large numbers by the 1870s, increased productivity by more than 500%. The sewing machine was an important productivity tool for mechanized shoe production.

With the widespread availability of machine tools, improved steam engines and inexpensive transportation provided by railroads, the machinery industry became the largest sector (by profit added) of the U. S. economy by the last quarter of the 19th century, leading to an industrial economy.

The first commercially successful glass bottle blowing machine was introduced in 1905. The machine, operated by a two-man crew working 12-hour shifts, could produce 17,280 bottles in 24 hours, compared to 2,880 bottles made a crew of six men and boys working in a shop for a day. The cost of making bottles by machine was 10 to 12 cents per gross compared to $1.80 per gross by the manual glassblowers and helpers.

Machine Tools

Machine tools, which cut, grind and shape metal parts, were another important mechanical innovation of the Industrial Revolution. Before machine tools it was prohibitively expensive to make precision parts, an essential requirement for many machines and interchangeable parts. Historically important machine tools are the screw-cutting lathe, milling machine and metal planer (metalworking), which all came into use between 1800 and 1840. However, around 1900, it was the combination of small electric motors, specialty steels and new cutting and grinding materials that allowed machine tools to mass-produce steel parts. Production of the Ford Model T required 32,000 machine tools.

Modern manufacturing began around 1900 when machines, aided by electric, hydraulic and pneumatic power, began to replace hand methods in industry. An early example is the Owens automatic glass bottle blowing machine, which reduced labor in making bottles by over 80%.

Mining

Large mining machines, such as steam shovels, appeared in the mid-nineteenth century, but were restricted to rails until the widespread introduction of continuous track and pneumatic tires in the late 19th and early 20th centuries. Until then much mining work was mostly done with pneumatic drills, jackhammers, picks and shovels.

Coal seam undercutting machines appeared around 1890 and were used for 75% of coal production by 1934. Coal loading was still being done manually with shovels around 1930, but mechanical pick up and loading machines were coming into use. The use of the coal boring machine improved productivity of sub-surface coal mining by a factor of three between 1949 and 1969.

There is currently a transition going under way from more labor-intensive methods of mining to more mechanization and even automated mining.

Mechanized Materials Handling

Bulk materials handling

P & H 4100 XPB cable loading shovel, a type of mobile crane

Unloading cotton c. 1900. Hydraulic cranes were in use in the U.K. for loading ships by the 1840s, but were little used in the U.S. Steam powered conveyors and cranes were used in the U.S. by the 1880s. In the early 20th century, electric operated cranes and motorized mobile loaders such as forklifts were used. Today non-bulk freight is containerized.

A U.S. airman operating a forklift. Pallets placed in rear of truck are moved around inside with a pallet jack (below). Where available pallets are loaded at loading docks which allow forklifts to drive on.

Dry bulk materials handling systems use a variety of stationary equipment such as conveyors, stackers, reclaimers and mobile equipment such as power shovels and loaders to handle high volumes of ores, coal, grains, sand, gravel, crushed stone, etc. Bulk materials handling systems are used at mines, for loading and unloading ships and at factories that process bulk materials into finished goods, such as steel and paper mills.

The handle on this pump jack is the lever for a hydraulic jack, which can easily lift loads up to 2-1/2 tonnes, depending on rating. Commonly used in warehouses and in retail stores.

Mechanical stokers for feeding coal to locomotives were in use in the 1920s. A completely mechanized and automated coal handling and stoking system was first used to feed pulverized coal to an electric utility boiler in 1921.

Liquids and gases are handled with centrifugal pumps and compressors, respectively.

Conversion to powered material handling increased during WW 1 as shortages of unskilled labor developed and unskilled wages rose relative to skilled labor.

A noteworthy use of conveyors was Oliver Evans's automatic flour mill built in 1785.

Around 1900 various types of conveyors (belt, slat, bucket, screw or auger), overhead cranes and industrial trucks began being used for handling materials and goods in various stages of production in factories.

A well known application of conveyors is Ford. Motor Co.'s assembly line (c. 1913), although Ford used various industrial trucks, overhead cranes, slides and whatever devices necessary to minimize labor in handling parts in various parts of the factory.

Cranes

Cranes are an ancient technology but they became widespread following the Industrial Revolution. Industrial cranes were used to handle heavy machinery at the Nasmyth, Gaskell and Company (Bridgewater foundry) in the late 1830s. Hydraulic powered cranes became widely used in the late 19th century, especially at British ports. Some cities, such as London, had public utility hydraulic service networks to power. Steam cranes were also used in the late 19th century. Electric cranes, especially the overhead type, were introduce in factories at the end of the 19th century. Steam cranes were usually restricted to rails. Continuous track (caterpillar tread) was developed in the late 19th century.

The important categories of cranes are:

- Overhead crane or bridge cranes-travel on a rail and have trolleys that move the hoist to any position inside the crane frame. Widely used in factories.

- Mobile crane Usually gasoline or diesel powered and travel on wheels for on or off road, rail or continuous track. They are widely used in construction, mining, excavation handling bulk materials.

- Fixed crane In a fixed position but can usually rotate full circle. The most familiar example is the tower crane used to erect tall buildings.

Palletization

Handling goods on pallets was a significant improvement over using hand trucks or carrying sacks or boxes by hand and greatly speeded up loading and unloading of trucks, rail cars and ships. Pallets can be handled with pallet jacks or forklift trucks which began being used in industry in the 1930s and became widespread by the 1950s. Loading docks built to architectural standards allow trucks or rail cars to load and unload at the same elevation as the warehouse floor.

Piggyback Rail

Piggyback is the transporting of trailers or entire trucks on rail cars, which is a more fuel efficient means of shipping and saves loading, unloading and sorting labor. Wagons had been carried on rail cars in the 19th century, with horses in separate cars. Trailers began being carried on rail cars in the U.S. in 1956. Piggyback was 1% of freight in 1958, rising to 15% in 1986.

Containerization

Either loading or unloading break bulk cargo on and off ships typically took several days. It was strenuous and somewhat dangerous work. Losses from damage and theft were high. The work was erratic and most longshoreman had a lot of unpaid idle time. Sorting and keeping track of break bulk cargo was also time consuming, and holding it in warehouses tied up capital.

Old style ports with warehouses were congested and many lacked efficient transportation infrastructure, adding to costs and delays in port.

By handling freight in standardized containers in compartmentalized ships, either loading or unloading could typically be accomplished in one day. Containers can be more efficiently filled than break bulk because containers can be stacked several high, doubling the freight capacity for a given size ship.

Loading and unloading labor for containers is a fraction of break bulk, and damage and theft are much lower. Also, many items shipped in containers require less packaging.

Containerization with small boxes was used in both world wars, particularly WW II, but became commercial in the late 1950s. Containerization left large numbers of warehouses at wharves in port cities vacant, freeing up land for other development.

Work Practices and Processes

Division of Labor

Before the factory system much production took place in the household, such as spinning and weaving, and was for household consumption. This was partly due to the lack of transportation infrastructures, especially in America.

Division of labor was practiced in antiquity but became increasingly specialized during the Industrial Revolution, so that instead of a shoemaker cutting out leather as part of the operation of making a shoe, a worker would do nothing but cut out leather. In Adam Smith's famous example of a pin factory, workers each doing a single task were far more productive than a craftsmen making an entire pin.

Starting before and continuing into the industrial revolution, much work was subcontracted under the putting out system (also called the domestic system) whereby work was done at home. Putting out work included spinning, weaving, leather cutting and, less commonly, specialty items such as firearms parts. Merchant capitalists or master craftsmen typically provided the materials and collected the work pieces, which were made into finished product in a central workshop.

Factory System

During the industrial revolution much production took place in workshops, which were typically located in the rear or upper level of the same building where the finished goods were sold. These workshops used tools and sometimes simple machinery, which was usually hand or animal powered. The master craftsman, foreman or merchant capitalist supervised the work and maintained quality. Workshops grew in size but were displaced by the factory system in the early 19th century. Under the factory system capitalists hired workers and provided the buildings, machinery and supplies and handled the sale of the finished products.

Interchangeable Parts

Changes to traditional work processes that were done after analyzing the work and making it more systematic greatly increased the productivity of labor and capital. This was the changeover from the European system of craftsmanship, where a craftsman made a whole item, to the American system of manufacturing which used special purpose machines and machine tools that made parts with precision to be interchangeable. The process took decades to perfect at great expense because interchangeable parts were more costly at first. Interchangeable parts were achieved by using fixtures to hold and precisely align parts being machined, jigs to guide the machine tools and gauges to measure critical dimensions of finished parts.

Scientific Management

Other work processes involved minimizing the amount of steps in doing individual tasks, such as bricklaying, by performing time and motion studies to determine the one best method, the system becoming known as Taylorism after Fredrick Winslow Taylor who is the best known developer of this method, which is also known as *scientific management* after his work *The Principles of Scientific Management*.

Standardization

Standardization and interchangeability are considered to be main reasons for U.S. exceptionality. Standardization was part of the change to interchangeable parts, but was also facilitated by the railroad industry and mass-produced goods. Railroad track gauge standardization and standards for rail cars allowed inter-connection of railroads. Railway time formalized time zones. Industrial standards included screw sizes and threads and later electrical standards. Shipping container standards were loosely adopted in the late 1960s and formally adopted ca. 1970. Today there are vast numbers of technical standards. Commercial standards includes such things as bed sizes. Architectural standards cover numerous dimensions including stairs, doors, counter heights and other designs to make buildings safe, functional and in some cases allow a degree of interchangeability.

Rationalized Factory Layout

Electrification allowed the placement of machinery such as machine tools in a systematic arrangement along the flow of the work. Electrification was a practical way to motorize conveyors to transfer parts and assemblies to workers, which was a key step leading to mass production and the assembly line.

Modern Business Management

Business administration, which includes management practices and accounting systems is another important form of work practices. As the size of businesses grew in the second half of the 19th century they began being organized by departments and managed by professional managers as opposed to being run by sole proprietors or partners.

Business administration as we know it was developed by railroads who had to keep up with trains, railcars, equipment, personnle and freight over large territories.

Modern business enterprise (MBE) is the organization and management of businesses, particularly large ones. MBE's employ professionals who use knowledge based techniques such areas as engineering, research and development, information technology, business administration, finance and accounting. MBE's typically benefit from economies of scale.

"Before railroad accounting we were moles burrowing in the dark." Andrew Carnegie

Continuous Production

Continuous production is a method by which a process operates without interruption for long periods, perhaps even years. Continuous production began with blast furnaces in ancient times and became popular with mechanized processes following the invention of the Fourdrinier paper machine during the Industrial Revolution, which was the inspiration for continuous rolling. It began being widely used in chemical and petroleum refining industries in the late nineteenth and early twentieth centuries. It was later applied to direct strip casting of steel and other metals.

Early steam engines did not supply power at a constant enough load for many continuous applications ranging from cotton spinning to rolling mills, restricting their power source to water. Advances in steam engines such as the Corliss steam engine and the development of control theory led to more constant engine speeds, which made steam power useful for sensitive tasks such as cotton spinning. AC motors, which run at constant speed even with load variations, were well suited to such processes.

Scientific Agriculture

Losses of agricultural products to spoilage, insects and rats contributed greatly to productivity. Much hay stored outdoors was lost to spoilage before indoor storage or some means of coverage became common. Pasteurization of milk allowed it to be shipped by railroad.

Keeping livestock indoors in winter reduces the amount of feed needed. Also, feeding chopped hay and ground grains, particularly corn (maize), was found to improve digestibility. The amount of feed required to produce a kg of live weight chicken fell from 5 in 1930 to 2 by the late 1990s and the time required fell from three months to six weeks.

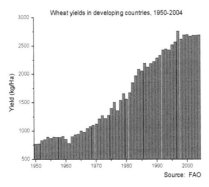

Wheat yields in developing countries, 1950 to 2004, kg/HA baseline 500. The steep rise in crop yields in the U.S. began in the 1940s. The percentage of growth was fastest in the early rapid growth stage. In developing countries maize yields are still rapidly rising.

The Green Revolution increased crop yields by a factor of 3 for soybeans and between 4 and 5 for corn (maize), wheat, rice and some other crops. Using data for corn (maize) in the U.S., yields increased about 1.7 bushels per acre from the early 1940s until the first decade of the 21st century when concern was being expressed about reaching limits of photosynthesis. Because of the constant nature of the yield increase, the annual percentage increase has declined from over 5% in the 1940s to 1% today, so while yields for a while outpaced population growth, yield growth now lags population growth.

High yields would not be possible without significant applications of fertilizer, particularly nitrogen fertilizer which was made affordable by the Haber-Bosch ammonia process. Nitrogen fertilizer is applied in many parts of Asia in amounts subject to diminishing returns, which however does still give a slight increase in yield. Crops in Africa are in general starved for NPK and much of the world's soils are deficient in zinc, which leads to deficiencies in humans.

The greatest period of agricultural productivity growth in the U.S. occurred from World War 2 until the 1970s.

Land is considered a form of capital, but otherwise has received little attention relative to its importance as a factor of productivity by modern economists, although it was important in classical economics. However, higher crop yields effectively multiplied the amount of land.

New Materials, Processes and De-materialization

Iron and Steel

The process of making cast iron was known before the 3rd century AD in China. Cast iron production reached Europe in the 14th century and Britain around 1500. Cast iron was useful for casting into pots and other implements, but was too brittle for making most tools. However, cast iron had a lower melting temperature than wrought iron and was much easier to make with primitive technology. Wrought iron was the material used for making many hardware items, tools and other implements. Before cast iron was made in Europe, wrought iron was made in small batches by the bloomery process, which was never used in China. Wrought iron could be made from cast iron more cheaply than it could be made with a bloomery.

The inexpensive process for making good quality wrought iron was puddling, which became widespread after 1800. Puddling involved stirring molten cast iron until small globs sufficiently decarburized to form globs of hot wrought iron that were then removed and hammered into shapes. Puddling was extremely labor-intensive. Puddling was used until the introduction of the Bessemer and open hearth processes in the mid and late 19th century, respectively.

Blister steel was made from wrought iron by packing wrought iron in charcoal and heating for several days.

The blister steel could be heated and hammered with wrought iron to make shear steel, which was used for cutting edges like scissors, knives and axes. Shear steel was of non uniform quality and a better process was needed for producing watch springs, a popular luxury item in the 18th century. The successful process was crucible steel, which was made by melting wrought iron and blister steel in a crucible.

Production of steel and other metals was hampered by the difficulty in producing sufficiently high temperatures for melting. An understanding of thermodynamic principles such as recapturing heat from flue gas by preheating combustion air, known as hot blast, resulted in much higher energy efficiency and higher temperatures. Preheated combustion air was used in iron production and in the open hearth furnace. In 1780, before the introduction of hot blast in 1829, it required seven times as much coke as the weight of the product pig iron. The hundredweight of coke per short ton of pig iron was 35 in 1900, falling to 13 in 1950. By 1970 the most efficient blast furnaces used 10 hundredweight of coke per short ton of pig iron.

Steel has much higher strength than wrought iron and allowed long span bridges, high rise buildings, automobiles and other items. Steel also made superior threaded fasteners (screws, nuts, bolts), nails, wire and other hardware items. Steel rails lasted over 10 times longer than wrought iron rails.

The Bessemer and open hearth processes were much more efficient than making steel by the puddling process because they used the carbon in the pig iron as a source of heat. The Bessemer (patented in 1855) and the Siemens-Martin (c. 1865) processes greatly reduced the cost of steel. By the end of the 19th century, Gilchirst-Thomas "basic" process had reduced production costs by 90% compared to the puddling process of the mid-century.

Today a variety of alloy steels are available that have superior properties for special applications like automobiles, pipelines and drill bits. High speed or tool steels, whose development began in the late 19th century, allowed machine tools to cut steel at much higher speeds. High speed steel and even harder materials were an essential component of mass production of automobiles.

Some of the most important specialty materials are steam turbine and gas turbine blades, which have to withstand extreme mechanical stress and high temperatures.

The size of blast furnaces grew greatly over the 20th century and innovations like additional heat recovery and pulverized coal, which displaced coke and increased energy efficiency.

Bessemer steel became brittle with age because nitrogen was introduced when air was blown in. The Bessemer process was also restricted to certain ores (low phosphate hematite). By the end of the 19th century the Bessemer process was displaced by the open hearth furnace (OHF). After World War II the OHF was displaced by the basic oxygen furnace (BOF), which used oxygen instead of air and required about 35–40 minutes to produce a batch of steel compared to 8 to 9 hours for the OHF. The BOF also was more energy efficient.

By 1913, 80% of steel was being made from molten pig iron directly from the blast furnace, eliminating the step of casting the "pigs" (ingots) and remelting.

The continuous wide strip rolling mill, developed by ARMCO in 1928, was most important development in steel industry during the inter-war years. Continuous wide strip rolling started with a with thick, coarse ingot. It produced a smoother sheet with more uniform thickness, which was better for stamping and gave a nice painted surface. It was good for automotive body steel and appliances. It used only a fraction of the labor of the discontinuous process, and was safer because it did not require continuous handling. Continuous rolling was made possible by improved sectional speed control.

After 1950 continuous casting contributed to productivity of converting steel to structural shapes by eliminating the intermittent step of making slabs, billets (square cross-section) or blooms (rectangular) which then usually have to be reheated before rolling into shapes. Thin slab casting, introduced in 1989, reduced labor to less than one hour per ton. Continuous thin slab casting and the BOF were the two most important productivity advancements in 20th-century steel making.

As a result of these innovations, between 1920 and 2000 labor requirements in the steel industry decreased by a factor of 1,000, from more than 3 worker-hours per tonne to just 0.003.

Sodium Carbonate (Soda Ash) and Related Chemicals

Sodium compounds: carbonate, bicarbonate and hydroxide are important industrial chemicals used in important products like making glass and soap. Until the invention of the Leblanc process in 1791, sodium carbonate was made, at high cost, from the ashes of seaweed and the plant barilla. The Leblanc process was replaced by the Solvay process beginning in the 1860s. With the widespread availability of inexpensive electricity, much sodium is produced along with chlorine by electro-chemical processes.

Cement

Cement is the binder for concrete, which is one of the most widely used construction materials today because of its low cost, versatility and durability. Portland cement, which was invented 1824-5, is made by calcining limestone and other naturally occurring minerals in a kiln. A great advance was the perfection of rotary cement kilns in the 1890s, the method still being used today. Reinforced concrete, which is suitable for structures, began being used in the early 20th century.

Paper

Paper was made one sheet at a time by hand until development of the Fourdrinier paper machine (c. 1801) which made a continuous sheet. Paper making was severely limited by the supply of cotton and linen rags from the time of the invention of the printing press until the development of wood pulp (c. 1850s)in response to a shortage of rags. The sulfite process for making wood pulp started operation in Sweden in 1874. Paper made from sulfite pulp had superior strength properties than the previously used ground wood pulp (c. 1840). The kraft (Swedish for *strong*) pulping process was commercialized in the 1930s. Pulping chemicals are recovered and internally recycled in the kraft process, also saving energy and reducing pollution. Kraft paperboard is the material that the outer layers of corrugated boxes are made of. Until Kraft corrugated boxes were available, packaging consisted of poor quality paper and paperboard boxes along with wood boxes and crates. Corrugated boxes require much less labor to manufacture than wooden boxes and offer good protection to their contents. Shipping containers reduce the need for packaging.

Rubber and Plastics

Vulcanized rubber made the pneumatic tire possible, which in turn enabled the development of on and off road vehicles as we know them. Synthetic rubber became important during the Second World War when supplies of natural rubber were cut off.

Rubber inspired a class of chemicals known as elastomers, some of which are used by themselves or in blends with rubber and other compounds for seals and gaskets, shock absorbing bumpers and a variety of other applications.

Plastics can be inexpensively made into everyday items and have significantly lowered the cost of a variety of goods including packaging, containers, parts and household piping.

Optical Fiber

Optical fiber began to replace copper wire in the telephone network during the 1980s. Optical fibers are very small diameter, allowing many to be bundled in a cable or conduit. Optical fiber is also an energy efficient means of transmitting signals.

Oil and Gas

Seismic exploration, beginning in the 1920s, uses reflected sound waves to map subsurface geology to help locate potential oil reservoirs. This was a great improvement over previous methods, which involved mostly luck and good knowledge of geology, although luck continued to be important in several major discoveries. Rotary drilling was a faster and more efficient way of drilling oil and water wells. It became popular after being used for the initial discovery of the East Texas field in 1930.

Hard Materials for Cutting

Numerous new hard materials were developed for cutting edges such as in machining. Mushet steel, which was developed in 1868, was a forerunner of High speed steel, which was developed by a team led by Fredrick Winslow Taylor at Bethlehem Steel Company around 1900. High speed steel held its hardness even when it became red hot. It was followed by a number of modern alloys.

From 1935 to 1955 machining cutting speeds increased from 120–200 ft/min to 1000 ft/min due to harder cutting edges, causing machining costs to fall by 75%.

One of the most important new hard materials for cutting is tungsten carbide.

Dematerialization

Dematerialization is the reduction of use of materials in manufacturing, construction, packaging or other uses. In the U.S. the quantity of raw materials per unit of output decreased approx 60% since 1900. In Japan the reduction has been 40% since 1973.

Dematerialization is made possible by substitution with better materials and by engineering to reduce weight while maintaining function. Modern examples are plastic beverage containers replacing glass and paperboard, plastic shrink wrap used in shipping and light weight plastic packing materials. Dematerialization has been occurring in the U. S. steel industry where the peak in consumption occurred in 1973 on both an absolute and per capita basis. At the same time, per capita steel consumption grew globally through outsourcing. Cumulative global GDP or wealth has grown in direct proportion to energy consumption since 1970, while Jevons paradox posits that efficiency improvement leads to increased energy consumption. Access to energy globally constrains dematerialization.

Communications

Telegraphy

The telegraph appeared around the beginning of the railroad era and railroads typically installed telegraph lines along their routes for communicating with the trains.

Teleprinters appeared in 1910 and had replaced between 80 and 90% of Morse code operators by 1929. It is estimated that one teletypist replaced 15 Morse code operators.

Telephone

The early use of telephones was primarily for business. Monthly service cost about one third of the average worker's earnings. The telephone along with trucks and the new road networks allowed businesses to reduce inventory sharply during the 1920s.

Telephone calls were handled by operators using switchboards until the automatic switchboard was introduced in 1892. By 1929, 31.9% of the Bell system was automatic.

Automatic telephone switching originally used electro-mechanical switches controlled by vacuum tube devices, which consumed a large amount of electricity. Call volume eventually grew so fast that it was feared the telephone system would consume all electricity production, prompting Bell Labs to begin research on the transistor.

Radio Frequency Transmission

After WWII microwave transmission began being used for long distance telephony and transmitting television programming to local stations for rebroadcast.

Fiber Optics

The diffusion of telephony to households was mature by the arrival of fiber optic communications in the late 1970s. Fiber optics greatly increased the transmission capacity of information over previous copper wires and further lowered the cost of long distance communication.

Communications Satellites

Communications satellites came into use in the 1960s and today carry a variety of information including credit card transaction data, radio, television and telephone calls. The Global Positioning System (GPS) operates on signals from satellites.

Facsimile (FAX)

Fax (short for facsimile) machines of various types had been in existence since the early 1900s but became widespread beginning in the mid-1970s.

Home Economics: Public Water Supply Household Gas Supply and Appliances

Before public water was supplied to households it was necessary for someone annually to haul up to 10,000 gallons of water to the average household.

Natural gas began being supplied to households in the late 19th century.

Household appliances followed household electrification in the 1920s, with consumers buying electric ranges, toasters, refrigerators and washing machines. As a result of appliances and convenience foods, time spent on meal preparation and clean up, laundry and cleaning decreased from 58 hours/week in 1900 to 18 hours/week by 1975. Less time spent on housework allowed more women to enter the labor force.

Automation, Process Control and Servomechanisms

Automation means automatic control, meaning a process is run with minimum operator intervention. Some of the various levels of automation are: mechanical methods, electrical relay, feedback control with a controller and computer control. Common applications of automation are for controlling temperature, flow and pressure. Automatic speed control is important in many industrial applications, especially in sectional drives, such as found in metal rolling and paper drying.

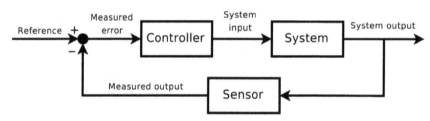

The concept of the feedback loop to control the dynamic behavior of the system: this is negative feedback, because the sensed value is subtracted from the desired value to create the error signal, which is processed by the controller, which provides proper corrective action. A typical example would be to control the opening of a valve to hold a liquid level in a tank. Process control is a widely used form of automation.

The earliest applications of process control were mechanisms that adjusted the gap between mill stones for grinding grain and for keeping windmills facing into the wind. The centrifugal governor used for adjusting the mill stones was copied by James Watt for controlling speed of steam engines in response to changes in heat load to the boiler; however, if the load on the engine changed the governor only held the speed steady at the new rate. It took much development work to achieve the degree of steadiness necessary to operate textile machinery. A mathematical analysis of control theory was first developed by James Clerk Maxwell. Control theory was developed to its "classical" form by the 1950s.

Factory electrification brought simple electrical controls such as ladder logic, whereby push buttons could be used to activate relays to engage motor starters. Other controls such as interlocks, timers and limit switches could be added to the circuit.

Today automation usually refers to feedback control. An example is cruise control on a car, which applies continuous correction when a sensor on the controlled variable (Speed in this example) deviates from a set-point and can respond in a corrective manner to hold the setting. Process control is the usual form of automation that allows industrial operations like oil refineries, steam plants generating electricity or paper mills to be run with a minimum of manpower, usually from a number of control rooms.

The need for instrumentation grew with the rapidly growing central electric power stations after the First World War. Instrumentation was also important for heat treating ovens, chemical plants and refineries. Common instrumentation was for measuring temperature, pressure or flow. Readings were typically recorded on circle charts or strip charts. Until the 1930s control was typically "open loop", meaning that it did not use feedback. Operators made various adjustments by such means as turning handles on valves. If done from a control room a message could be sent to an operator in the plant by color coded light, letting him know whether to increase or decrease whatever was being controlled. The signal lights were operated by a switchboard, which soon became automated. Automatic control became possible with the feedback controller, which sensed the measured variable, measured the deviation from the setpoint and perhaps the rate of change and time weighted amount of deviation, compared that with the setpoint and automatically applied a calculated adjustment. A stand-alone controller may use a combination of mechanical, pneumatic, hydraulic or electronic analogs to manipulate the controlled device. The tendency was to use electronic controls after these were developed, but today the tendency is to use a computer to replace individual controllers.

By the late 1930s feedback control was gaining widespread use. Feedback control was an important technology for continuous production.

Automation of the telephone system allowed dialing local numbers instead of having calls placed through an operator. Further automation allowed callers to place long distance calls by direct dial. Eventually almost all operators were replaced with automation.

Machine tools were automated with Numerical control (NC) in the 1950s. This soon evolved into computerized numerical control (CNC).

Servomechanisms are commonly position or speed control devices that use feedback. Understanding of these devices is covered in control theory. Control theory was successfully applied to steering ships in the 1890s, but after meeting with personnel resistance it was not widely implemented for that application until after the First World War. Servomechanisms are extremely important in providing automatic stability control for airplanes and in a wide variety of industrial applications.

A set of six-axis robots used for welding. Robots are commonly used for hazardous jobs like paint spraying, and for repetitive jobs requiring high precision such as welding and the assembly and soldering of electronics like car radios.

Industrial robots were used on a limited scale from the 1960s but began their rapid growth phase in the mid-1980s after the widespread availability of microprocessors used for their control. By 2000 there were over 700,000 robots world-wide.

Computers, Semiconductors, Data Processing and Information Technology

Unit Record Equipment

Early IBM tabulating machine. Common applications were accounts receivable, payroll and billing.

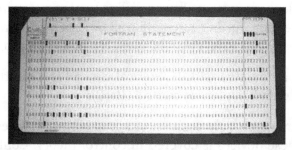

Card from a Fortran program: Z(1) = Y + W(1). The punched card carried over from tabulating machines to stored program computers before being replaced by terminal input and magnetic storage.

Early electric data processing was done by running punched cards through tabulating machines, the holes in the cards allowing electrical contact to increment electronic counters. Tabulating machines were in a category called unit record equipment, through which the flow of punched cards was arranged in a program-like sequence to allow sophisticated data processing. Unit record equipment was widely used before the introduction of computers.

The usefulness of tabulating machines was demonstrated by compiling the 1890 U.S. census, allowing the census to be processed in less than a year and with great labor savings compared to the estimated 13 years by the previous manual method.

Stored Program Computers

The first digital computers were more productive than tabulating machines, but not by a great amount. Early computers used thousands of vacuum tubes (thermionic valves) which used a lot of electricity and constantly needed replacing. By the 1950s the vacuum tubes were replaced by transistors which were much more reliable and used relatively little electricity. By the 1960s thousands of transistors and other electronic components could be manufactured on a silicon semiconductor wafer as integrated circuits, which are universally used in today's computers.

Computers used paper tape and punched cards for data and programming input until the 1980s when it was still common to receive monthly utility bills printed on a punched card that was returned with the customer's payment.

In 1973 IBM introduced point of sale (POS) terminals in which electronic cash registers were networked to the store mainframe computer. By the 1980s bar code readers were added. These technologies automated inventory management. Wal-Mart was an early adopter of POS. The Bureau of Labor Statistics estimated that bar code scanners at checkout increased ringing speed by 30% and reduced labor requirements of cashiers and baggers by 10-15%.

Data storage became better organized after the development of relational database software that allowed data to be stored in different tables. For example, a theoretical airline may have numerous tables such as: airplanes, employees, maintenance contractors, caterers, flights, airports, payments, tickets, etc. each containing a narrower set of more specific information than would a flat file, such as a spreadsheet. These tables are related by common data fields called *keys*. Data can be retrieved in various specific configurations by posing a *query* without having to pull up a whole table. This, for example, makes it easy to find a passenger's seat assignment by a variety of means such as ticket number or name, and provide only the *queried* information.

Since the mid-1990s, interactive web pages have allowed users to access various servers over Internet to engage in e-commerce such as online shopping, paying bills, trading stocks, managing bank accounts and renewing auto registrations. This is the ultimate form of back office automation because the transaction information is transferred directly to the database.

Computers also greatly increased productivity of the communications sector, especially in areas like the elimination of telephone operators. In engineering, computers replaced manual drafting with CAD, with a 500% average increase in a draftsman's output. Software was developed for calculations used in designing electronic circuits, stress analysis, heat and material balances. Process simulation software has been developed for both steady state and dynamic simulation, the latter able to give the user a very similar experience to operating a real process like a refinery or paper mill, allowing the user to optimize the process or experiment with process modifications.

Automated teller machines (ATM's) became popular in recent decades and self checkout at retailers appeared in the 1990s.

The Airline Reservations System and banking are areas where computers are practically essential. Modern military systems also rely on computers.

In 1959 Texaco's Port Arthur refinery became the first chemical plant to use digital process control.

Computers did not revolutionize manufacturing because automation, in the form of control systems, had already been in existence for decades, although computers did allow more sophisticated control, which led to improved product quality and process optimization.

Long Term Decline in Productivity Growth

"The years 1929-1941 were, in the aggregate, the most technologically progressive of any comparable period in U.S. economic history." Alexander J. Field

"As industrialization has proceeded, its effects, relatively speaking, have become less, not more, revolutionary"...."There has, in effect, been a general progression in industrial commodities from a deficiency to a surplus of capital relative to internal investments". Alan Sweezy, 1943

U.S. productivity growth has been in long term decline since the early 1970s, with the exception of a 1996–2004 spike caused by an acceleration of Moore's law semiconductor innovation. Part of the early decline was attributed to increased governmental regulation since the 1960s, including stricter environmental regulations. Part of the decline in productivity growth is due to exhaustion of opportunities, especially as the traditionally high productivity sectors decline in size. Robert J. Gordon considered productivity to be "One big wave" that crested and is now receding to a lower level, while M. King Hubbert called the phenomenon of the great productivity gains preceding the Great Depression a "one time event."

Because of reduced population growth in the U.S. and a peaking of productivity growth, sustained U.S. GDP growth has never returned to the 4% plus rates of the pre-World War 1 decades.

The computer and computer-like semiconductor devices used in automation are the most significant productivity improving technologies developed in the final decades of the twentieth century; however, their contribution to overall productivity growth was disappointing. Most of the productivity growth occurred in the new industry computer and related industries. Economist Robert J. Gordon is among those who questioned whether computers lived up to the great innovations of the past, such as electrification. This issue is known as the Productivity paradox. Gordon's (2013) analysis of productivity in the U.S. gives two possible surges in growth, one during 1891–1972 and the second in 1996–2004 due to the acceleration in Moore's law-related technological innovation.

Improvements in productivity affected the relative sizes of various economic sectors by reducing prices and employment. Agricultural productivity released labor at a time when manufacturing was growing. Manufacturing productivity growth peaked with factory electrification and automation, but still remains significant. However, as the relative size of the manufacturing sector shrank the government and service sectors, which have low productivity growth, grew.

Improvement in Living Standards

Chronic hunger and malnutrition were the norm for the majority of the population of the world including England and France, until the latter part of the 19th century. Until about 1750, in large part due to malnutrition, life expectancy in France was about 35 years, and only slightly higher in England. The U.S. population of the time was adequately fed, were much taller and had life expectancies of 45–50 years.

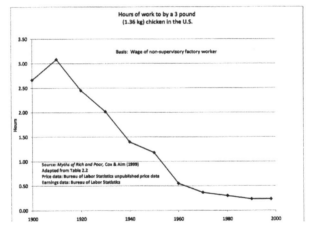

An hour's work in 1998 bought 11 times as much chicken as in 1900. Many consumer items show similar declines in terms of work time.

The gains in standards of living have been accomplished largely through increases in productivity. In the U.S. the amount of personal consumption that could be bought with one hour of work was about $3.00 in 1900 and increased to about $22 by 1990, measured in 2010 dollars. For comparison, a U. S. worker today earns more (in terms of buying power) working for ten minutes than subsistence workers, such as the English mill workers that Fredrick Engels wrote about in 1844, earned in a 12-hour day.

Decline in Work Week

As a result of productivity the work week declined considerably over the 19th century. By the 1920s the average work week in the U.S. was 49 hours, but the work week was reduced to 40 hours (after which overtime premium was applied) as part of the National Industrial Recovery Act of 1933.

Laser Beam Machining

Laser beam machining (LBM) is a non-traditional subtractive manufacturing process, a form of machining, in which a laser is directed towards the work piece for machining. This process uses thermal energy to remove material from metallic or nonmetallic surfaces. The laser is focused onto the surface to be worked and the thermal energy of the laser is transferred to the surface, heating and melting or vaporizing the material. Laser beam machining is best suited for brittle materials with low conductivity, but can be used on most materials.

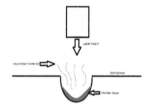

A visual of how laser beam machining works

Types of Lasers

There are many different types of lasers including gas, solid states lasers, and excimer.

In gas lasers, an electric current is liberated from a gas to generate a consistent light. Some of the most commonly used gases consist of; He-Ne, Ar, and CO_2. Fundamentally, these gases act as a pumping medium to ensure that the necessary population inversion is attained.

Solid state lasers are designed by doping a rare element into various host materials. Unlike in gas lasers, solid state lasers are pumped optically by flash lamps or arch lamps. Ruby is one of the frequently used host materials in this type of laser. A Ruby Laser is a type of the solid state laser whose laser medium is a synthetic ruby crystal. These ruby lasers generate deep red light pulses of a millisecond pulse length and a wavelength of about 694.3 nm. The synthetic ruby rod is optically pumped using a xenon flashtube before it is used as an active laser medium.

YAG is an abbreviation for yttrium aluminum garnet which are crystals that are used for solid-state lasers while Nd:YAG refers to neodymium-doped yttrium aluminum garnet crystals that are used in the solid-state lasers as the laser medium. The Nd:YAG lasers emit a wavelength of light waves with high energy. Nd:glass is neodymium–doped gain media made of either silicate or phosphate materials that are used in fiber laser.

In excimer lasers, the state is different than in solid state or gas lasers. The device utilizes a combination of reactive and inert gases to produce a beam. This machine is sometimes known as an ultraviolet chemical laser.

Cutting Depth

The cutting depth of a laser is directly proportional to the quotient obtained by dividing the power of the laser beam by the product of the cutting velocity and the diameter of the laser beam spot.

$$t \propto P v d ,$$

where t is the depth of cut, P is the laser beam power, v is the cutting velocity, and d is the laser beam spot diameter.

The depth of the cut is also influenced by the workpiece material. The material's reflectivity, density, specific heat, and melting point temperature all contribute to the lasers ability to cut the workpiece.

The following table shows the ability of different lasers to cut different materials:

material	laser: 10.6	wavelength (micrometer) Nd:YAG laser: 1.06
ceramics	well	poorly
plywood	very well	fairly well
polycarbonate	well	fairly well
polyethylene	very well	fairly well
Perspex	very well	fairly well
Titanium	well	well
Gold	not possible	well
Copper	poorly	well
Aluminium	well	well
stainless steel		very well
construction steel		very well

Applications

Lasers can be used for welding, cladding, marking, surface treatment, drilling, and cutting among other manufacturing processes. It is used in the automobile, shipbuilding, aerospace, steel, electronics, and medical industries for precision machining of complex parts. Laser welding is advantageous in that it can weld at speeds of up to 100 mm/s as well as the ability to weld dissimilar metals. Laser cladding is used to coat cheep or weak parts with a harder material in order to improve the surface quality. Drilling and cutting with lasers is advantageous in that there is little to no wear on the cutting tool as there is no contact to cause damage. Milling with a laser is a three dimensional process that requires two lasers, but drastically cuts costs of machining parts. Lasers can be used to change the surface properties of a workpiece.

Laser beam machining can also be used in conjunction with traditional machining methods. By focusing the laser ahead of a cutting tool the material to be cut will be softened and made easier to remove, reducing cost of production and wear on the tool while increasing tool life.

The appliance of laser beam machining varies depending on the industry. In heavy manufacturing laser beam machining is used for cladding and drilling, spot and seam welding among others. In light manufacturing the machine is used to engrave and to drill other metals. In the electronic industry laser beam machining is used for wire stripping and skiving of circuits. In the medical industry it is used for cosmetic surgery and hair removal.

Advantages

Since the rays of a laser beam are monochromatic and parallel it can be focused to a very small diameter and can produce energy as high as 100 MW of energy for a square millimeter of area. It is especially suited to making accurately placed holes.

Laser beam machining has the ability to engrave or cut nearly all materials, where traditional cutting methods may fall short. There are several types of lasers, and each have different uses. For instance, materials that cannot be cut by a gas laser may be cut by an excimer laser.

The cost of maintaining lasers is moderately low due to the low rate of wear and tear, as there is no physical contact between the tool and the workpiece.

The machining provided by laser beams is high precision, and most of these processes do not require additional finishing.

Laser beams can be paired with gases to help the cutting process be more efficient, help minimize oxidization of surfaces, and/or keep the workpiece surface free from melted or vaporized material.

Disadvantages

The initial cost of acquiring a laser beam is moderately high. There are many accessories that aid in the machining process, and as most of these accessories are as important as the laser beam itself the startup cost of machining is raised further.

Handling and maintaining the machining requires highly trained individuals. Operating the laser beam is comparatively technical, and services from an expert may be required.

Laser beams are not designed to produce mass metal processes. For this reason production is always slow, especially when the metal processes involve a lot of cutting.

Laser beam machining consumes a lot of energy.

Deep cuts are difficult with workpieces with high melting points and usually cause a taper.

Rapid Prototyping

Rapid prototyping is a group of techniques used to quickly fabricate a scale model of a physical part or assembly using three-dimensional computer aided design (CAD) data. Construction of the part or assembly is usually done using 3D printing or "additive layer manufacturing" technology.

A rapid prototyping machine using selective laser sintering (SLS)

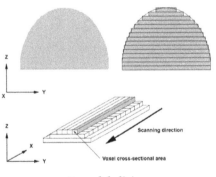

3D model slicing

The first methods for rapid prototyping became available in the late 1980s and were used to produce models and prototype parts. Today, they are used for a wide range of applications and are used to manufacture production-quality parts in relatively small numbers if desired without the typical unfavorable short-run economics. This economy has encouraged online service bureaus. Historical surveys of RP technology start with discussions of simulacra production techniques used by 19th-century sculptors. Some modern sculptors use the progeny technology to produce exhibitions. The ability to reproduce designs from a dataset has given rise to issues of rights, as it is now possible to interpolate volumetric data from one-dimensional images.

As with CNC subtractive methods, the computer-aided-design - computer-aided manufacturing CAD-CAM workflow in the traditional Rapid Prototyping process starts with the creation of geometric data, either as a 3D solid using a CAD workstation, or

2D slices using a scanning device. For RP this data must represent a valid geometric model; namely, one whose boundary surfaces enclose a finite volume, contain no holes exposing the interior,and do not fold back on themselves. In other words, the object must have an "inside." The model is valid if for each point in 3D space the computer can determine uniquely whether that point lies inside, on, or outside the boundary surface of the model. CAD post-processors will approximate the application vendors' internal CAD geometric forms (e.g., B-splines) with a simplified mathematical form, which in turn is expressed in a specified data format which is a common feature in Additive Manufacturing: STL (stereolithography) a de facto standard for transferring solid geometric models to SFF machines. To obtain the necessary motion control trajectories to drive the actual SFF, Rapid Prototyping, 3D Printing or Additive Manufacturing mechanism, the prepared geometric model is typically sliced into layers, and the slices are scanned into lines [producing a "2D drawing" used to generate trajectory as in CNC`s toolpath], mimicking in reverse the layer-to-layer physical building process.

Application Areas

Rapid Prototyping and Production Automotive Spareparts

Electric cars can be built and tested in one year with 3D production systems.

Software Development

Rapid prototyping is commonly applied in software engineering to try out new business models and application architectures.

History

In the 1970s, Joseph Henry Condon and others at Bell Labs developed the Unix Circuit Design System (UCDS), automating the laborious and error-prone task of manually converting drawings to fabricate circuit boards for the purposes of research and development.

In the 1980s U.S. policy makers and industrial managers were forced to take note that America's dominance in the field of machine tool manufacturing evaporated, in what was named the machine tool crisis. Numerous projects sought to counter these trends in the traditional CNC CAM area, which had begun in the US. Later when Rapid Prototyping Systems moved out of labs to be commercialized it was recognized that developments were already international and U.S. rapid prototyping companies would not have the luxury of letting a lead slip away. The National Science Foundation was an umbrella for the National Aeronautics and Space Administration (NASA), the US Department of Energy, the US Department of Commerce NIST, the US Department of Defense, Defense Advanced Research Projects Agency (DARPA), and the Office of Naval Research coordinated studies to inform strategic planners in their deliberations.

One such report was the 1997 *Rapid Prototyping in Europe and Japan Panel Report* in which Joseph J. Beaman founder of DTM Corporation [DTM RapidTool pictured] provides a historical perspective:

The technologies referred to as Solid Freeform Fabrication are what we recognize today as Rapid Prototyping, 3D Printing or Additive Manufacturing: Swainson (1977), Schwerzel (1984) worked on polymerization of a photosensitive polymer at the intersection of two computer controlled laser beams. Ciraud (1972) considered magnetostatic or electrostatic deposition with electron beam, laser or plasma for sintered surface cladding. These were all proposed but it is unknown if working machines were built. Hideo Kodama of Nagoya Municipal Industrial Research Institute was the first to publish an account of a solid model fabricated using a photopolymer rapid prototyping system (1981). Even at that early date the technology was seen as having a place in manufacturing practice. A low resolution, low strength output had value in design verification, mould making, production jigs and other areas. Outputs have steadily advanced toward higher specification uses.

Innovations are constantly being sought,to improve speed and the ability to cope with mass production applications. A dramatic development which RP shares with related CNC areas is the freeware open-sourcing of high level applications which constitute an entire CAD-CAM toolchain. This has created a community of low res device manufacturers. Hobbyists have even made forays into more demanding laser-effected device designs.

Techniques

- 3D printing (3DP)

- Ballistic particle manufacturing (BPM)

- Directed light fabrication (DLF)

- Direct-shell production casting (DSPC)

- Fused deposition modeling (FDM)

- Laminated object manufacturing (LOM)

- Shape deposition manufacturing (SDM) (and Mold SDM)

- Solid ground curing (SGC)

- Stereolithography (SL)

- Selective laser sintering (SLS)

- Agile software development

- Cloud computing

Electrical Discharge Machining

Electrical discharge machining (EDM), sometimes colloquially also referred to as spark machining, spark eroding, burning, die sinking, wire burning or wire erosion, is a manufacturing process whereby a desired shape is obtained by using electrical discharges (sparks). Material is removed from the workpiece by a series of rapidly recurring current discharges between two electrodes, separated by a dielectric liquid and subject to an electric voltage. One of the electrodes is called the tool-electrode, or simply the "tool" or "electrode," while the other is called the workpiece-electrode, or "workpiece." The process depends upon the tool and workpiece not making actual contact.

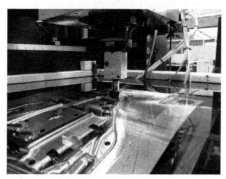

An electrical discharge machine

When the voltage between the two electrodes is increased, the intensity of the electric field in the volume between the electrodes becomes greater than the strength of the dielectric (at least in some places), which breaks down, allowing current to flow between the two electrodes. This phenomenon is the same as the breakdown of a capacitor (condenser). As a result, material is removed from the electrodes. Once the current stops (or is stopped, depending on the type of generator), new liquid dielectric is usually conveyed into the inter-electrode volume, enabling the solid particles (debris) to be carried away and the insulating properties of the dielectric to be restored. Adding new liquid dielectric in the inter-electrode volume is commonly referred to as "flushing." Also, after a current flow, the difference of potential between the electrodes is restored to what it was before the breakdown, so that a new liquid dielectric breakdown can occur.

History

The erosive effect of electrical discharges was first noted in 1770 by English physicist Joseph Priestley.

Die-sink Edm

Two Russian scientists, B. R. Butinzky and N. I. Lazarenko, were tasked in 1943 to investigate ways of preventing the erosion of tungsten electrical contacts due to sparking. They failed in this task but found that the erosion was more precisely controlled if the electrodes were immersed in a dielectric fluid. This led them to invent an EDM machine used for working difficult-to-machine materials such as tungsten. The Lazarenkos' machine is known as an R-C-type machine, after the resistor–capacitor circuit (RC circuit) used to charge the electrodes.

Simultaneously but independently, an American team, Harold Stark, Victor Harding, and Jack Beaver, developed an EDM machine for removing broken drills and taps from aluminium castings. Initially constructing their machines from feeble electric-etching tools, they were not very successful. But more powerful sparking units, combined with automatic spark repetition and fluid replacement with an electromagnetic interrupter arrangement produced practical machines. Stark, Harding, and Beaver's machines were able to produce 60 sparks per second. Later machines based on their design used vacuum tube circuits that were able to produce thousands of sparks per second, significantly increasing the speed of cutting.

Wire-cut EDM

The wire-cut type of machine arose in the 1960s for the purpose of making tools (dies) from hardened steel. The tool electrode in wire EDM is simply a wire. To avoid the erosion of material from the wire causing it to break, the wire is wound between two spools so that the active part of the wire is constantly changing. The earliest numerical controlled (NC) machines were conversions of punched-tape vertical milling machines. The first commercially available NC machine built as a wire-cut EDM machine was manufactured in the USSR in 1967. Machines that could optically follow lines on a master drawing were developed by David H. Dulebohn's group in the 1960s at Andrew Engineering Company for milling and grinding machines. Master drawings were later produced by computer numerical controlled (CNC) plotters for greater accuracy. A wire-cut EDM machine using the CNC drawing plotter and optical line follower techniques was produced in 1974. Dulebohn later used the same plotter CNC program to directly control the EDM machine, and the first CNC EDM machine was produced in 1976.

Generalities

Electrical discharge machining is a machining method primarily used for hard metals or those that would be very difficult to machine with traditional techniques. EDM typically works with materials that are electrically conductive, although methods for machining insulating ceramics with EDM have also been proposed. EDM can cut intricate contours or cavities in pre-hardened steel without the need for heat treatment to soften and re-harden them. This method can be used with any other metal or metal alloy such as titanium, hastelloy, kovar, and inconel. Also, applications of this process to shape polycrystalline diamond tools have been reported.

EDM is often included in the "non-traditional" or "non-conventional" group of machining methods together with processes such as electrochemical machining (ECM), water jet cutting (WJ, AWJ), laser cutting and opposite to the "conventional" group (turning, milling, grinding, drilling and any other process whose material removal mechanism is essentially based on mechanical forces).

Ideally, EDM can be seen as a series of breakdown and restoration of the liquid dielectric in-between the electrodes. However, caution should be exerted in considering such a statement because it is an idealized model of the process, introduced to describe the fundamental ideas underlying the process. Yet, any practical application involves many aspects that may also need to be considered. For instance, the removal of the debris from the inter-electrode volume is likely to be always partial. Thus the electrical properties of the dielectric in the inter-electrodes volume can be different from their nominal values and can even vary with time. The inter-electrode distance, often also referred to as spark-gap, is the end result of the control algorithms of the specific machine used. The control of such a distance appears logically to be central to this process. Also, not all of the current between the dielectric is of the ideal type described above: the spark-gap can be short-circuited by the debris. The control system of the electrode may fail to react quickly enough to prevent the two electrodes (tool and workpiece) from coming into contact, with a consequent short circuit. This is unwanted because a short circuit contributes to material removal differently from the ideal case. The flushing action can be inadequate to restore the insulating properties of the dielectric so that the current always happens in the point of the inter-electrode volume (this is referred to as arcing), with a consequent unwanted change of shape (damage) of the tool-electrode and workpiece. Ultimately, a description of this process in a suitable way for the specific purpose at hand is what makes the EDM area such a rich field for further investigation and research.

To obtain a specific geometry, the EDM tool is guided along the desired path very close to the work; ideally it should not touch the workpiece, although in reality this may happen due to the performance of the specific motion control in use. In this way, a large number of current discharges (colloquially also called sparks) happen, each contributing to the removal of material from both tool and workpiece, where small craters are formed. The size of the craters is a function of the technological parameters set for the specific job at hand. They can be with typical dimensions ranging from the nanoscale (in micro-EDM operations) to some hundreds of micrometers in roughing conditions.

The presence of these small craters on the tool results in the gradual erosion of the electrode. This erosion of the tool-electrode is also referred to as wear. Strategies are needed to counteract the detrimental effect of the wear on the geometry of the workpiece. One possibility is that of continuously replacing the tool-electrode during a machining operation. This is what happens if a continuously replaced wire is used as electrode. In this case, the correspondent EDM process is also called wire EDM. The tool-electrode can also be used in such a way that only a small portion of it is actually engaged in the machining process and this portion is changed on a regular basis. This is, for instance, the case when using a rotating disk as a tool-electrode. The corresponding process is often also referred to as EDM grinding.

A further strategy consists in using a set of electrodes with different sizes and shapes during the same EDM operation. This is often referred to as multiple electrode strategy, and is most common when the tool electrode replicates in negative the wanted shape and is advanced towards the blank along a single direction, usually the vertical direction (i.e. z-axis). This resembles the sink of the tool into the dielectric liquid in which the workpiece is immersed, so, not surprisingly, it is often referred to as die-sinking EDM (also called conventional EDM and ram EDM). The corresponding machines are often called sinker EDM. Usually, the electrodes of this type have quite complex forms. If the final geometry is obtained using a usually simple-shaped electrode which is moved along several directions and is possibly also subject to rotations, often the term EDM milling is used.

In any case, the severity of the wear is strictly dependent on the technological parameters used in the operation (for instance: polarity, maximum current, open circuit voltage). For example, in micro-EDM, also known as μ-EDM, these parameters are usually set at values which generates severe wear. Therefore, wear is a major problem in that area.

The problem of wear to graphite electrodes is being addressed. In one approach, a digital generator, controllable within milliseconds, reverses polarity as electro-erosion takes place. That produces an effect similar to electroplating that continuously deposits the eroded graphite back on the electrode. In another method, a so-called "Zero Wear" circuit reduces how often the discharge starts and stops, keeping it on for as long a time as possible.

Definition of The Technological Parameters

Difficulties have been encountered in the definition of the technological parameters that drive the process.

Two broad categories of generators, also known as power supplies, are in use on EDM machines commercially available: the group based on RC circuits and the group based on transistor controlled pulses.

In the first category, the main parameters to choose from at setup time are the resistance(s) of the resistor(s) and the capacitance(s) of the capacitor(s). In an ideal condition these quantities would affect the maximum current delivered in a discharge which is expected to be associated with the charge accumulated on the capacitors at a certain moment in time. Little control, however, is expected over the time duration of the discharge, which is likely to depend on the actual spark-gap conditions (size and pollution) at the moment of the discharge. The RC circuit generator can allow the user to obtain short time durations of the discharges more easily than the pulse-controlled generator, although this advantage is diminishing with the development of new electronic components. Also, the open circuit voltage (i.e. the voltage between the electrodes when the dielectric is not yet broken) can be identified as steady state voltage of the RC circuit.

In generators based on transistor control, the user is usually able to deliver a train of pulses of voltage to the electrodes. Each pulse can be controlled in shape, for instance, quasi-rectangular. In particular, the time between two consecutive pulses and the duration of each pulse can be set. The amplitude of each pulse constitutes the open circuit voltage. Thus, the maximum duration of discharge is equal to the duration of a pulse of voltage in the train. Two pulses of current are then expected not to occur for a duration equal or larger than the time interval between two consecutive pulses of voltage.

The maximum current during a discharge that the generator delivers can also be controlled. Because other sorts of generators may also be used by different machine builders, the parameters that may actually be set on a particular machine will depend on the generator manufacturer. The details of the generators and control systems on their machines are not always easily available to their user. This is a barrier to describing unequivocally the technological parameters of the EDM process. Moreover, the parameters affecting the phenomena occurring between tool and electrode are also related to the controller of the motion of the electrodes.

A framework to define and measure the electrical parameters during an EDM operation directly on inter-electrode volume with an oscilloscope external to the machine has been recently proposed by Ferri *et al.* These authors conducted their research in the field of μ-EDM, but the same approach can be used in any EDM operation. This would enable the user to estimate directly the electrical parameters that affect their operations without relying upon machine manufacturer's claims. When machining different materials in the same setup conditions, the actual electrical parameters of the process are significantly different.

Material Removal Mechanism

The first serious attempt of providing a physical explanation of the material removal during electric discharge machining is perhaps that of Van Dijck. Van Dijck presented a thermal model together with a computational simulation to explain the phenomena between the electrodes during electric discharge machining. However, as Van Dijck himself admitted in his study, the number of assumptions made to overcome the lack of experimental data at that time was quite significant.

Further models of what occurs during electric discharge machining in terms of heat transfer were developed in the late eighties and early nineties, including an investigation at Texas A&M University with the support of AGIE, now Agiecharmilles. It resulted in three scholarly papers: the first presenting a thermal model of material removal on the cathode, the second presenting a thermal model for the erosion occurring on the anode and the third introducing a model describing the plasma channel formed during the passage of the discharge current through the dielectric liquid. Validation of these models is supported by experimental data provided by AGIE.

These models give the most authoritative support for the claim that EDM is a thermal process, removing material from the two electrodes because of melting and/or vaporization, along with pressure dynamics established in the spark-gap by the collapsing of the plasma channel. However, for small discharge energies the models are inadequate to explain the experimental data. All these models hinge on a number of assumptions from such disparate research areas as submarine explosions, discharges in gases, and failure of transformers, so it is not surprising that alternative models have been proposed more recently in the literature trying to explain the EDM process.

Among these, the model from Singh and Ghosh reconnects the removal of material from the electrode to the presence of an electrical force on the surface of the electrode that could mechanically remove material and create the craters. This would be possible because the material on the surface has altered mechanical properties due to an increased temperature caused by the passage of electric current. The authors' simulations showed how they might explain EDM better than a thermal model (melting and/or evaporation), especially for small discharge energies, which are typically used in μ-EDM and in finishing operations.

Given the many available models, it appears that the material removal mechanism in EDM is not yet well understood and that further investigation is necessary to clarify it, especially considering the lack of experimental scientific evidence to build and validate the current EDM models. This explains an increased current research effort in related experimental techniques.

Types

Sinker EDM

Sinker EDM allowed quick production of 614 uniform injectors for the J-2 rocket engine, six of which were needed for each trip to the moon.

Sinker EDM, also called cavity type EDM or volume EDM, consists of an electrode and workpiece submerged in an insulating liquid such as, more typically, oil or, less frequently, other dielectric fluids. The electrode and workpiece are connected to a suitable power supply. The power supply generates an electrical potential between the two parts. As the electrode approaches the workpiece, dielectric breakdown occurs in the fluid, forming a plasma channel, and a small spark jumps.

These sparks usually strike one at a time, because it is very unlikely that different locations in the inter-electrode space have the identical local electrical characteristics which would enable a spark to occur simultaneously in all such locations. These sparks happen in huge numbers at seemingly random locations between the electrode and the workpiece. As the base metal is eroded, and the spark gap subsequently increased, the electrode is lowered automatically by the machine so that the process can continue uninterrupted. Several hundred thousand sparks occur per second, with the actual duty cycle carefully controlled by the setup parameters. These controlling cycles are sometimes known as "on time" and "off time", which are more formally defined in the literature.

The on time setting determines the length or duration of the spark. Hence, a longer on time produces a deeper cavity for that spark and all subsequent sparks for that cycle, creating a rougher finish on the workpiece. The reverse is true for a shorter on time. Off time is the period of time that one spark is replaced another. A longer off time, for example, allows the flushing of dielectric fluid through a nozzle to clean out the eroded debris, thereby avoiding a short circuit. These settings can be maintained in microseconds. The typical part geometry is a complex 3D shape, often with small or odd shaped angles. Vertical, orbital, vectorial, directional, helical, conical, rotational, spin and indexing machining cycles are also used.

Wire EDM

CNC Wire-cut EDM machine

In *wire electrical discharge machining* (WEDM), also known as *wire-cut EDM* and *wire cutting*, a thin single-strand metal wire, usually brass, is fed through the workpiece, submerged in a tank of dielectric fluid, typically deionized water. Wire-cut EDM is typically used to cut plates as thick as 300mm and to make punches, tools, and dies from hard metals that are difficult to machine with other methods. The wire, which is constantly fed from a spool, is held between upper and lower diamond guides. The guides, usually CNC-controlled, move in the x–y plane. On most machines, the upper guide can also move independently in the z–u–v axis, giving rise to the ability to cut tapered and transitioning shapes (circle on the bottom, square at the top for example). The upper guide can control axis movements in x–y–u–v–i–j–k–l–. This allows the wire-cut EDM to be programmed to cut very intricate and delicate shapes. The upper and lower diamond guides are usually accurate to 0.004 mm (0.16 mils), and can have a cutting path or *kerf* as small as 0.021 mm (0.83 mils) using Ø 0.02 mm (0.79 mils) wire, though the average cutting kerf that achieves the best economic cost and machining time is 0.335 mm (13.2 mils) using Ø 0.25 mm (9.8 mils) brass wire. The reason that the cutting width is greater than the width of the wire is because sparking occurs from the sides of the wire to the work piece, causing erosion. This "overcut" is necessary, for many applications it is adequately predictable and therefore can be compensated for (for instance in micro-EDM this is not often the case). Spools of wire are long — an 8 kg spool of 0.25 mm wire is just over 19 kilometers in length. Wire diameter can be as small as 20 µm (0.79 mils) and the geometry precision is not far from ± 1 µm (0.039 mils). The wire-cut process uses water as its dielectric fluid, controlling its resistivity and other electrical properties with filters and de-ionizer units. The water flushes the cut debris away from the cutting zone. Flushing is an important factor in determining the maximum feed rate for a given material thickness. Along with tighter tolerances, multi axis EDM wire-cutting machining centers have added features such as multi heads for cutting two parts at the same time, controls for preventing wire breakage, automatic self-threading features in case of wire breakage, and programmable machining strategies to optimize the operation. Wire-cutting EDM is commonly used when low residual stresses are desired, because it does not require high cutting forces for removal of material. If the energy/power per pulse is relatively low (as in finishing operations), little change in the mechanical properties of a material is expected due to these low residual stresses, although material that hasn't been stress-relieved can distort in the machining process. The work piece may undergo a significant thermal cycle, its severity depending on the technological parameters used. Such thermal cycles may cause formation of a recast layer on the part and residual tensile stresses on the work piece. If machining takes place after heat treatment, dimensional accuracy will not be affected by heat treat distortion.

Applications

Prototype Production

The EDM process is most widely used by the mold-making tool and die industries, but is becoming a common method of making prototype and production parts, especially in the aerospace, automobile and electronics industries in which production quantities are relatively low. In sinker EDM, a graphite, copper tungsten or pure copper electrode is machined into the desired (negative) shape and fed into the workpiece on the end of a vertical ram.

Coinage Die Making

For the creation of dies for producing jewelry and badges, or blanking and piercing (through use of a pancake die) by the coinage (stamping) process, the positive master may be made from sterling silver, since (with appropriate machine settings) the master is significantly eroded and is used only once. The resultant negative die is then hardened and used in a drop hammer to produce stamped flats from cutout sheet blanks of bronze, silver, or low proof gold alloy. For badges these flats may be further shaped to a curved surface by another die. This type of EDM is usually performed submerged in an oil-based dielectric. The finished object may be further refined by hard (glass) or soft (paint) enameling and/or electroplated with pure gold or nickel. Softer materials such as silver may be hand engraved as a refinement.

Master at top, badge die workpiece at bottom, oil jets at left (oil has been drained). Initial flat stamping will be "dapped" to give a curved surface.

EDM control panel (Hansvedt machine). Machine may be adjusted for a refined surface (electropolish) at end of process.

Small Hole Drilling

Small hole drilling EDM is used in a variety of applications.

On wire-cut EDM machines, small hole drilling EDM is used to make a through hole in a workpiece in through which to thread the wire for the wire-cut EDM operation. A separate EDM head specifically for small hole drilling is mounted on a wire-cut machine and allows large hardened plates to have finished parts eroded from them as needed and without pre-drilling.

A turbine blade with internal cooling as applied in the high-pressure turbine.

Small hole drilling EDM machines.

Small hole EDM is used to drill rows of holes into the leading and trailing edges of turbine blades used in jet engines. Gas flow through these small holes allows the engines to use higher temperatures than otherwise possible. The high-temperature, very hard, single crystal alloys employed in these blades makes conventional machining of these holes with high aspect ratio extremely difficult, if not impossible.

Small hole EDM is also used to create microscopic orifices for fuel system components, spinnerets for synthetic fibers such as rayon, and other applications.

There are also stand-alone small hole drilling EDM machines with an x–y axis also known as a super drill or *hole popper* that can machine blind or through holes. EDM drills bore holes with a long brass or copper tube electrode that rotates in a chuck with a constant flow of distilled or deionized water flowing through the electrode as a flushing agent and dielectric. The electrode tubes operate like the wire in wire-cut EDM machines, having a spark gap and wear rate. Some small-hole drilling EDMs are able to drill through 100 mm of soft or through hardened steel in less than 10 seconds, averaging 50% to 80% wear rate. Holes of 0.3 mm to 6.1 mm can be achieved in this drilling operation. Brass electrodes are easier to machine but are not recommended for wire-cut operations due to eroded brass particles causing "brass on brass" wire breakage, therefore copper is recommended.

Metal Disintegration Machining

Several manufacturers produce MDM machines for the specific purpose of removing broken tools (drill bits, taps, bolts and studs) from work pieces. In this application, the process is termed "metal disintegration machining" or MDM. The metal disintegration process removes only the center out of the tap, bolt or stud leaving the hole intact and allowing a part to be reclaimed.

Closed Loop Manufacturing

Closed loop manufacturing can improve the accuracy and reduce the tool costs

Advantages and Disadvantages

Advantages of EDM include machining of:

- Complex shapes that would otherwise be difficult to produce with conventional cutting tools.

- Extremely hard material to very close tolerances.

- Very small work pieces where conventional cutting tools may damage the part from excess cutting tool pressure.

- There is no direct contact between tool and work piece. Therefore, delicate sections and weak materials can be machined without perceivable distortion.

- A good surface finish can be obtained; a very good surface may be obtained by redundant finishing paths.

- Very fine holes can be attained.

- Tapered holes may be produced.

Disadvantages of EDM include:

- The slow rate of material removal.

- Potential fire hazard associated with use of combustible oil based dielectrics.

- The additional time and cost used for creating electrodes for ram/sinker EDM.

- Reproducing sharp corners on the workpiece is difficult due to electrode wear.

- Specific power consumption is very high.

- Power consumption is high.

- "Overcut" is formed.

- Excessive tool wear occurs during machining.

- Electrically non-conductive materials can be machined only with specific set-up of the process.

References

- Boothroyd, Geoffrey; Knight, Winston A. (1989), Fundamentals of machining and machine tools (2nd ed.), Marcel Dekker, pp. 478–9, ISBN 978-0-8247-7852-1.

- Todd, Robert H.; Allen, Dell K.; Alting, Leo (1994), Manufacturing Processes Reference Guide, Industrial Press Inc., pp. 2–5, ISBN 0-8311-3049-0.

- Atack, Jeremy; Passell, Peter (1994). A New Economic View of American History. New York: W.W. Norton and Co. p. 156. ISBN 0-393-96315-2.

- Rosen, William (2012). The Most Powerful Idea in the World: A Story of Steam, Industry and Invention. University Of Chicago Press. p. 137. ISBN 978-0226726342.

- Robert U. Ayres and Benjamin Warr, The Economic Growth Engine: How useful work creates material prosperity, 2009. ISBN 978-1-84844-182-8.

- Wells, David A. (1891). Recent Economic Changes and Their Effect on Production and Distribution of Wealth and Well-Being of Society. New York: D. Appleton and Co. p. 416. ISBN 0-543-72474-3.

- Williams, Trevor I. (1993). A Short History of Twentieth Century Technology. USA: Oxford University Press. p. 30. ISBN 978-0198581598.

- Paepke, C. Owen (1992). The Evolution of Progress: The End of Economic Growth and the Beginning of Human Transformation. New York, Toronto: Random House. p. 109. ISBN 0-679-41582-3.

- McNeil, Ian (1990). An Encyclopedia of the History of Technology. London: Routledge. ISBN 0-415-14792-1.

- Field, Alexander J. (2011). A Great Leap Forward: 1930s Depression and U.S. Economic Growth. New Haven, London: Yale University Press. ISBN 978-0-300-15109-1.

- Clark, Gregory (2007). A Farewell to Alms: A Brief Economic History of the World. Princeton University Press. p. 286. ISBN 978-0-691-12135-2.

- Bealer, Alex W.. The tools that built America. Mineola, NY: Dover Publications, 2004. 12-13. ISBN 0486437531.

- Thomson, Ross (1989). The Path to Mechanized Shoe Production in the United States. University of North Carolina Press. ISBN 978-0807818671.

Weaving Machinery: An Overview

Loom is used in weaving cloth and tapestry; the purpose of any loom is to basically hold the warp threads in order to simplify the weaving of the threads. Lancashire loom, rapier loom and reed are some of the topics discussed in the following section. The chapter is an overview of the subject matter incorporating all the major aspects of weaving machinery.

Loom

A loom is a device used to weave cloth and tapestry. The basic purpose of any loom is to hold the warp threads under tension to facilitate the interweaving of the weft threads. The precise shape of the loom and its mechanics may vary, but the basic function is the same.

A foot-treadle operated Hattersley & Sons, Domestic Loom, built under license in 1893, in Keighley, Yorkshire.

Etymology

The word "loom" is derived from the Old English *"geloma"* formed from ge-(perfective prefix) and *"loma"*, a root of unknown origin; this meant utensil or tool or machine of any kind. In 1404 it was used to mean a machine to enable weaving thread into cloth. By 1838 it had gained the meaning of a machine for interlacing thread.

Weaving

Weaving is done by intersecting the longitudinal threads, the warp, i.e. "that which is thrown across", with the transverse threads, the weft, i.e. "that which is woven".

Weaving demonstration on a 1830 handloom in the weaving museum in Leiden

The major components of the loom are the warp beam, heddles, harnesses or shafts (as few as two, four is common, sixteen not unheard of), shuttle, reed and takeup roll. In the loom, yarn processing includes shedding, picking, battening and taking-up operations. These are the principal motions.

- Shedding. Shedding is the raising of part of the warp yarn to form a shed (the vertical space between the raised and unraised warp yarns), through which the filling yarn, carried by the shuttle, can be inserted. On the modern loom, simple and intricate shedding operations are performed automatically by the heddle or heald frame, also known as a harness. This is a rectangular frame to which a series of wires, called heddles or healds, are attached. The yarns are passed through the eye holes of the heddles, which hang vertically from the harnesses. The weave pattern determines which harness controls which warp yarns, and the number of harnesses used depends on the complexity of the weave. Two common methods of controlling the heddles are dobbies and a Jacquard Head.

Shuttles

- Picking. As the harnesses raise the heddles or healds, which raise the warp yarns, the shed is created. The filling yarn is inserted through the shed by a small carrier device called a shuttle. The shuttle is normally pointed at each end to allow passage through the shed. In a traditional shuttle loom, the filling yarn is wound onto a quill, which in turn is mounted in the shuttle. The filling yarn emerges through a hole in the shuttle as it moves across the loom. A single crossing of the shuttle from one side of the loom to the other is known as a pick. As the shuttle moves back and forth across the shed, it weaves an edge, or selvage, on each side of the fabric to prevent the fabric from raveling.

- Battening. Between the heddles and the takeup roll, the warp threads pass through another frame called the reed (which resembles a comb). The portion of the fabric that has already been formed but not yet rolled up on the takeup roll is called the fell. After the shuttle moves across the loom laying down the fill yarn, the weaver uses the reed to press (or batten) each filling yarn against the fell. Conventional shuttle looms can operate at speeds of about 150 to 160 picks per minute.

There are two secondary motions, because with each weaving operation the newly constructed fabric must be wound on a cloth beam. This process is called taking up. At the same time, the warp yarns must be let off or released from the warp beams. To become fully automatic, a loom needs a tertiary motion, the filling stop motion. This will brake the loom, if the weft thread breaks. An automatic loom requires 0.125 hp to 0.5 hp to operate.

Types of Looms

Back Strap Loom

A back strap loom with a shed-rod.

A simple loom which has its roots in ancient civilizations consists of two sticks or bars between which the warps are stretched. One bar is attached to a fixed object, and the other to the weaver usually by means of a strap around the back. On traditional looms, the two main sheds are operated by means of a shed roll over which one set of warps pass, and continuous string heddles which encase each of the warps in the other set. The weaver leans back and uses his or her body weight to tension the loom. To open the shed controlled by the string heddles, the weaver relaxes tension on the warps and raises the heddles. The other shed is usually opened by simply drawing the shed roll toward the weaver. Both simple and complex textiles can be woven on this loom. Width is limited to how far the weaver can reach from side to side to pass the shuttle. Warp faced textiles, often decorated with intricate pick-up patterns woven in complementary and supplementary warp techniques are woven by indigenous peoples today around the world. They produce such things as belts, ponchos, bags, hatbands and carrying cloths. Supplementary weft patterning and brocading is practiced in many regions. Balanced weaves are also possible on the backstrap loom. Today, commercially produced backstrap loom kits often include a rigid heddle.

Warp-weighted Loom

The warp-weighted loom is a vertical loom that may have originated in the Neolithic period. The earliest evidence of warp-weighted looms comes from sites belonging to the Starčevo culture in modern Serbia and Hungary and from late Neolithic sites in Switzerland. This loom was used in Ancient Greece, and spread north and west throughout Europe thereafter. Its defining characteristic is hanging weights (loom weights) which keep bundles of the warp threads taut. Frequently, extra warp thread is wound around the weights. When a weaver has reached the bottom of the available warp, the completed section can be rolled around the top beam, and additional lengths of warp threads can be unwound from the weights to continue. This frees the weaver from vertical size constraints.

Drawloom

A drawloom is a hand-loom for weaving figured cloth. In a drawloom, a "figure harness" is used to control each warp thread separately. A drawloom requires two operators, the weaver and an assistant called a "drawboy" to manage the figure harness.

Handloom

A handloom is a simple machine used for weaving. In a wooden vertical-shaft looms, the heddles are fixed in place in the shaft. The warp threads pass alternately through a heddle, and through a space between the heddles (the shed), so that raising the shaft raises half the threads (those passing through the heddles), and lowering the shaft lowers the same threads — the threads passing through the spaces between the heddles remain in place. This was a great discovery in that era.

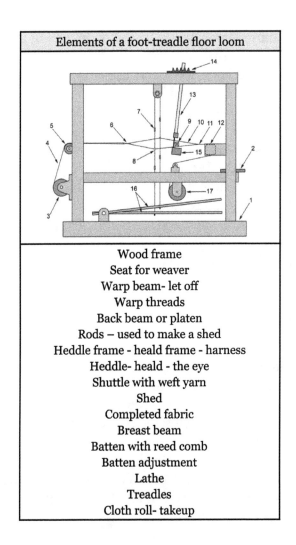

Elements of a foot-treadle floor loom

Wood frame
Seat for weaver
Warp beam- let off
Warp threads
Back beam or platen
Rods – used to make a shed
Heddle frame - heald frame - harness
Heddle- heald - the eye
Shuttle with weft yarn
Shed
Completed fabric
Breast beam
Batten with reed comb
Batten adjustment
Lathe
Treadles
Cloth roll- takeup

Flying Shuttle

Hand weavers could only weave a cloth as wide as their armspan. If cloth needed to be wider, two people would do the task (often this would be an adult with a child). John Kay (1704–1779) patented the flying shuttle in 1733. The weaver held a picking stick that was attached by cords to a device at both ends of the shed. With a flick of the wrist, one cord was pulled and the shuttle was propelled through the shed to the other end with considerable force, speed and efficiency. A flick in the opposite direction and the shuttle was propelled back. A single weaver had control of this motion but the flying shuttle could weave much wider fabric than an arm's length at much greater speeds than had been achieved with the hand thrown shuttle.

The *flying shuttle* was one of the key developments in weaving that helped fuel the Industrial Revolution. The whole picking motion no longer relied on manual skill and it was just a matter of time before it could be powered.

Haute-lisse and Basse-lisse Looms

Looms used for weaving traditional tapestry are classified as *haute-lisse* looms, where the warp is suspended vertically between two rolls, and the *basse-lisse* looms, where the warp extends horizontally between the rolls.

Ribbon Weaving

Traditional Looms

Several other types of hand looms exist, including the simple frame loom, pit loom, free-standing loom, and the pegged loom. Each of these can be constructed, and provide work and income in developing societies.

Power Looms

Two Lancashire looms in the Queen Street Mill weaving shed, Burnley

A 1939 loom working at the Mueller Cloth Mill museum in Euskirchen, Germany.

Edmund Cartwright built and patented a power loom in 1785, and it was this that was adopted by the nascent cotton industry in England. The silk loom made by Jacques Vaucanson in 1745 operated on the same principles but was not developed further. The invention of the flying shuttle by John Kay was critical to the development of a commercially successful power loom. Cartwright's loom was impractical but the ideas behind it were developed by numerous inventors in the Manchester area of England where, by 1818, there were 32 factories containing 5,732 looms.

Horrocks loom was viable, but it was the Roberts Loom in 1830 that marked the turning point. Incremental changes to the three motions continued to be made. The problems of sizing, stop-motions, consistent take-up, and a temple to maintain the width remained. In 1841, Kenworthy and Bullough produced the Lancashire Loom which was self-acting or semi-automatic. This enables a youngster to run six looms at the same time. Thus, for simple calicos, the power loom became more economical to run than the hand loom – with complex patterning that used a dobby or Jacquard head, jobs were still put out to handloom weavers until the 1870s. Incremental changes were made such as the Dickinson Loom, culminating in the Keighley-born inventor Northrop, who was working for the Draper Corporation in Hopedale producing the fully automatic Northrop Loom. This loom recharged the shuttle when the pirn was empty. The Draper E and X models became the leading products from 1909. They were challenged by synthetic fibres such as rayon.

From 1942 the faster and more efficient shuttleless Sulzer looms and the rapier looms were introduced. Modern industrial looms can weave at 2,000 weft insertions per minute.

Weft Insertion

A Picanol rapier loom

Different types of looms are most often defined by the way that the weft, or pick, is inserted into the warp. Many advances in weft insertion have been made in order to make manufactured cloth more cost effective. There are five main types of weft insertion and they are as follows:

- Shuttle: The first-ever powered looms were shuttle-type looms. Spools of weft are unravelled as the shuttle travels across the shed. This is very similar to projectile methods of weaving, except that the weft spool is stored on the shuttle. These looms are considered obsolete in modern industrial fabric manufacturing because they can only reach a maximum of 300 picks per minute.

- Air jet: An air-jet loom uses short quick bursts of compressed air to propel the weft through the shed in order to complete the weave. Air jets are the fastest traditional method of weaving in modern manufacturing and they are able to achieve up to 1,500 picks per minute. However, the amounts of compressed air required to run these looms, as well as the complexity in the way the air jets are positioned, make them more costly than other looms.

- Water jet: Water-jet looms use the same principle as air-jet looms, but they take advantage of pressurized water to propel the weft. The advantage of this type of weaving is that water power is cheaper where water is directly available on site. Picks per minute can reach as high as 1,000.

- Rapier loom: This type of weaving is very versatile, in that rapier looms can weave using a large variety of threads. There are several types of rapiers, but they all use a hook system attached to a rod or metal band to pass the pick across the shed. These machines regularly reach 700 picks per minute in normal production.

- Projectile: Projectile looms utilize an object that is propelled across the shed, usually by spring power, and is guided across the width of the cloth by a series of reeds. The projectile is then removed from the weft fibre and it is returned to the opposite side of the machine so it can get reused. Multiple projectiles are in use in order to increase the pick speed. Maximum speeds on these machines can be as high as 1,050 ppm.

Shedding

Dobby Looms

A dobby loom is a type of floor loom that controls the whole warp threads using a dobby head. Dobby is a corruption of "draw boy" which refers to the weaver's helpers who used to control the warp thread by pulling on draw threads. A dobby loom is an alternative to a treadle loom, where multiple heddles (shafts) were controlled by foot treadles – one for each heddle.

Jacquard Looms

The Jacquard loom is a mechanical loom, invented by Joseph Marie Jacquard in 1801, which simplifies the process of manufacturing textiles with complex patterns such as brocade, damask and matelasse. The loom is controlled by punched cards with punched holes, each row of which corresponds to one row of the design. Multiple rows of holes are punched on each card and the many cards that compose the design of the textile are strung together in order. It is based on earlier inventions by the Frenchmen Basile Bouchon (1725), Jean Baptiste Falcon (1728) and Jacques Vaucanson (1740) To call it a loom is a misnomer, a Jacquard head could be attached to a power loom or a hand loom, the head controlling which warp thread was raised during shedding. Multiple shuttles could be used to control the colour of the weft during picking.

Hand operated Jacquard looms in the Textile Department of the Strzemiński Academy of Fine Arts in Łódź, Poland.

Following the pattern, holes are punched in the appropriate places on a jacquard card.

Battening on a jacquard loom in Łódź.

Circular Looms

A circular loom is used to create a seamless tube of fabric for products such as hosiery, sacks, clothing, fabric hose (such as fire hose) and the like. Circular looms can be small jigs used for Circular knitting or large high-speed machines for modern garments. Modern circular looms use up to ten shuttles driven from below in a circular motion by electromagnets for the weft yarns, and cams to control the warp threads. The warps rise and fall with each shuttle passage, unlike the common practice of lifting all of them at once.

Lancashire Loom

The Lancashire Loom was a semi-automatic power loom invented by James Bullough and William Kenworthy in 1842. Although it is self-acting, it has to be stopped to recharge empty shuttles. It was the mainstay of the Lancashire cotton industry for a century.

Two Lancashire looms

John Bullough

John Bullough (1800–68) was from Accrington, often described as a simple-minded West-houghton weaver. Originally a handloom weaver, unlike others of his trade Bullough embraced new developments such as Edmund Cartwright's power loom (1785). While colleagues were busy rejecting new devices such as in the power-loom riots that broke out in Lancashire in 1826, Bullough improved his own loom by inventing various components, including the "self-acting temple" that kept the woven cloth at its correct width, and a loose reed that allowed the lathe to back away on encountering a shuttle trapped in the warp. Bullough also invented a simple but effective warning device which rang a bell every time a warp thread broke on his loom. Bullough moved to Blackburn and worked with William Kenworthy at Brookhouse Mills, with whom he applied his inventions to develop an improved power loom that later became known as the "Lancashire Loom". He was forced to quit Blackburn, for fear of angry handloom weavers. He later settled in Accrington to form Howard & Bullough in partnership with John Howard at Globe Works, alongside the Leeds-Liverpool Canal in Accrington. Here he invented the slasher, which founded the company's success. He was one of the country's largest manufacturers. At the height of the business the Globe works employed almost 6000 workers and covered 52 acres (210,000 m²). 75% of production was exported. Howard and Bullough became part of the Textile Machinery Makers Limited group, which were bought out by Platt, and in 1991 the company name changed to Platt Saco Lowell. The Globe works closed in 1993.

The Loom

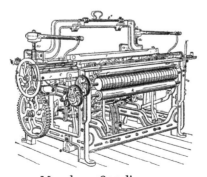

Marsdens:1892 diagram

1894 Lancashire Looms (Harling & Todd) still in use at Queen Street Mill, Burnley.

From 1830 there had been a series of incremental improvements to the basic Roberts Loom.

- Richard Roberts 1830, Roberts Loom. These improvements were a geared take up wheel and tappets to operate multiple heddles

- Stanford, Pritchard and Wilkinson – patented a method to stop on the break of weft or warp. It was not used.

- William Dickinson of Blackburn Blackburn Loom the modern overpick loom

There now appear a series of useful improvements that are contained in patents for useless devices

- Hornby, Kenworthy and Bullough of Blackburn 1834 – the vibrating or fly reed

- John Ramsden and Richard Holt of Todmorden 1834 – a new automatic weft stopping motion

- John Bullough of Blackburn 1835 – improved automatic weft stopping motion and taking up and letting off arrangements

- Andrew Parkinson 1836 – improved stretcher (temple).

- William Kenworthy and James Bullough 1841 – trough and roller temple (became the standard), A simple stop-motion.

At this point the loom has become fully automatic, this is the Kenworthy and Bullough Lancashire Loom. The Cartwight loom weaver could work one loom at 120–130 picks per minute- with a Kenworthy and Bullough's Lancashire Loom, a weaver can run up to six looms working at 220–260 picks per minute- thus giving 12 times more throughput. The power loom is now referred to as "a perfect machine", it produced textile of a better quality than the hand weaver for less cost. An economic success. Other improvements were the

- John Bullough 1842 – the loose reed, which doubled the operating speed

- John Sellers 1845 – Burnley Brake, a loom brake

- Blackburn 1852 – Dickinson Loom Modern overpick- or side pick using the cone and bowl that substituted the lever pick. Invented in Dickinson's mill.

Movements

The three primary movements of a loom are shedding, picking, and beating-up.

- *Shedding*: The operation of dividing the warp into two lines, so that the shuttle can pass between these lines. There are two general kinds of sheds: "open" and "closed". Open Shed-The warp threads are moved when the pattern requires it-from one line to the other. Closed Shed-The warp threads are all placed level in one line after each pick.

- *Picking*: The operation of projecting the shuttle from side to side of the loom through the division in the warp threads. This is done by the over-pick or underpick motions. The overpick is suitable for quick-running looms, whereas the underpick is best for heavy or slow looms.

- *Beating-up*: The third primary movement of the loom when making cloth, and is the action of the reed as it drives each pick of weft to the fell of the cloth.

Economics

The principal advantage of the Lancashire loom was that it was semi-automatic, when a warp thread broke the weaver was notified. When the shuttle ran out of thread, the machine stopped. An operative thus could work 4 or more looms whereas previously they could only work a single loom. Indeed, the term *A Four Loom Weaver* was used to describe the operatives. Labour cost was quartered. In some mills an operative would operate 6 or even 8 looms, although that was governed by the thread being used. By 1900, the loom was challenged by the Northrop Loom, which was fully automatic and could be worked in larger numbers. The Northrop was suitable for coarse thread but for fine cotton, the Lancashire loom was still preferred. By 1914, Northop looms made up 40% of looms in American mills, but in the United Kingdom labour costs were not as significant and they only supplied 2% of the British market.

Rapier Loom

A rapier loom is a shuttleless weaving loom in which the filling yarn is carried through the shed of warp yarns to the other side of the loom by finger-like carriers called rapiers.

A stationary package of yarn is used to supply the weft yarns in the rapier machine. One end of a rapier, a rod or steel tape, carries the weft yarn. The other end of the rapier is connected to the control system. The rapier moves across the width of the fabric, carrying the weft yarn across through the shed to the opposite side. The rapier is then retracted, leaving the new pick in place.

In some versions of the loom, two rapiers are used, each half the width of the fabric in size. One rapier carries the yarn to the centre of the shed, where the opposing rapier picks up the yarn and carries it the remainder of the way across the shed. The double rapier is used more frequently than the single rapier due to its increased pick insertion speed and ability to weave wider widths of fabric.

The housing for the rapiers must take up as much space as the width of the machine. To overcome this problem, looms with flexible rapiers have been devised. The flexible rapier can be coiled as it is withdrawn, therefore requiring less storage space. If, however, the rapier is too stiff then it will not coil; If it is too flexible, it will buckle. Rigid and flexible rapier machines operate at speeds ranging from about 200 to 260 ppm, using up to 1,300 metres of weft yarn every minute. They have a noise level similar to that of modern projectile looms. They can produce a wide variety of fabrics ranging from muslin to drapery and upholstery materials.

Newer rapier machines are built with two distinct weaving areas for two separate fabrics. On such machines, one rapier picks up the yarn from the centre, between the two fabrics, and carries it across one weaving area; as it finishes laying that pick, the opposite end of the rapier picks up another yarn from the centre, and the rapier moves in the other direction to lay a pick for the second weaving area, on the other half of the machine. The above figure shows the action on a single width of fabric for a single rigid rapier system, a double rigid rapier system, and a double flexible rapier system .

Rapier machines weave more rapidly than most shuttle machines but more slowly than most other projectile machines. An important advantage of rapier machines is their flexibility, which permits the laying of picks of different colours. They also weave yarns of any type of fibre and can weave fabrics up to 110 inches in width without modification.

History of The Rapier Loom

The development of the rapier loom began in 1844, when John Smith of Salford was granted a patent on a loom design that eliminated the shuttle typical of earlier models of looms. Subsequent patents were taken out by Phillippe and Maurice in 1855, W.S. Laycock in 1869, and W. Glover in 1874, with rigid rapiers being perfected by O. Hallensleben in 1899. The main breakthrough came in 1922 when John Gabler invented the principle of loop transfer in the middle of the shed. Flexible rapiers of the type used today were proposed in 1925 by the Spanish inventor R.G. Moya, while R. Dewas introduced the idea of grasping the weft at its tip by the giver or a carrier rapier and transferring it to the taker or a receiver in the middle of the shed. It was not until the 1950s and 1960s that rapier weaving became fully commercialized, with loom technology developing rapidly.

Reed (Weaving)

A reed is part of a loom, and resembles a comb. It is used to push the weft yarn securely into place as it is woven, it also separates the warp threads and holds them in their positions, keeping them untangled, and guides the shuttle as it moves across the loom. It consists of a frame with lots of vertical slits. The reed is securely held by the beater. Floor looms and mechanized looms both use a beater with a reed, whereas Inkle weaving and tablet weaving do not use reeds.

The reed is the part in the beater where the warp threads go through.

History

Modern reeds are made by placing flattened strips of wire (made of carbon or stainless steel) between two half round ribs of wood, and binding the whole together with tarred string. Historically reeds were made of reed or cane, however modern reeds are made of metal wires. John Kay in 1738 first used flattened iron or brass wire, and the change was quickly adopted. Previously the cane was split by pressing it against a spindle that had knives radiating out of it at the appropriate distance apart. The split cane was then bound between the ribs of wood in the same manner as the wire is now.

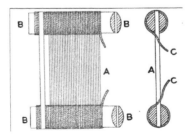

A: wires or dents
B: wooden ribs
C: tarred cord

The wire is flattened to a uniform thickness by passing it between rollers, straightened, given rounded edges and smoothed. The final step is to cut the wire to the correct length and insert it into position. The tarred cord that binds the reed together is wrapped around each set of wooden ribs and between the dents to hold the ribs together.

The length of the metal wire varies depending on the type of fabric and the type of loom being used. For a machine powered cotton loom, the metal wires are commonly 3.5 inches (89 mm) long. For hand powered floor looms, around 4 inches (100 mm) is common.

Dents

Both the wires and the slots in the reed are known as dents (namely, teeth). The warp threads pass through the dents after going through the heddles and before becoming woven cloth. The number of dents per inch (or per cm or per 10 cm) indicates the number of gaps per linear width, the number of the warp thread ends by weaving width determines the fineness of the cloth. One or more warp threads may go through each dent. The number of warp threads that go through each dent depends on the warp, and it is possible that the number of threads in each dent is not constant for a whole warp. The number of threads per dent might not be constant if the weaver alternates 2 and three threads per dent, in order to get a number of ends per inch that is 2.5 times the number of dents per inch, or if the thickness of the warp threads were to change at that point, and the fabric to have a thicker or thinner section.

A reed on end

One thread per dent is most common for coarse work, for finer work (20 or more ends per inch) two or more threads are put through each dent. Threads can be doubled in every other space, so that a reed with 10 dents per inch could give 15 ends per inch, or 20 if the threads were simply doubled. Also, threads can be put in every other dent so as to make a cloth with 6 ends per inch from a reed with 12 dents per inch. Putting more than one thread through each dent reduces friction and the number of reeds that one weaver needs, and is used in weaving mills. If too many threads are put through one dent there may be reed marks left in the fabric, especially in linen and cotton.

For cotton fabrics reeds typically have between 6 and 90 dents per inch. When the reed has a very high number of dents per inch, sometimes there are actually two rows of wires which are offset. This is to keep loose fibers from twisting and blocking the shed.

Interchangeability

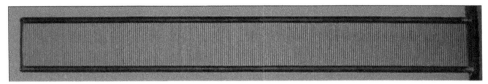

A reed with 5 dents per inch, separate from the loom

In order to weave many different patterns on the same loom different reeds are useful. Modern floor looms and several other types of loom use interchangeable reeds, with a different number of dents per inch. The most common sizes available for the hand-weaver are 6, 8, 10, 12, or 15, though more are made, and sizes between 5 and 24 are not uncommon. The finer the reed, the more dents per inch, and, in general, the more ends per inch in the final woven fabric. This is because by beating the weft into place the reed determines the distance between threads, or groups of threads. Having interchangeable reeds allows a weaver to do many different types of weaving on the same loom. By changing the reed the weaver can move from weaving fine fabric to coarse fabric without needing a different loom.

Reeds also come in different widths. The width of the reed determines the maximum width of the warp.

Sleying The Reed

Sleying is the term used for pulling the warp threads through the loom, which happens in the warping process (the process of putting a warp on the loom). Sleying is done by inserting a *reed hook* through the reed, hooking the warp threads and then pulling them through the dent. The warp threads are taken in the order they come from the heddles, so as to avoid crossing threads. If the threads cross, the shed will not open correctly when the weaving is begun.

References

- Cartwright, Wendy (2007). Weave. Murdoch Books. p. 10. ISBN 1-74045-978-4. Retrieved July 6, 2009.

- Black, Mary E. (1957). New Key to Weaving. Milwaukee: The Bruce Publishing Company. pp. 35–38. ISBN 0-02-511140-X.

- "Rapier Looms - Rapier Looms Weaving and Flexible Rapier Looms". www.rapierloom.in. Retrieved 2016-07-07.

Essential Aspects of Machine Tool

A machine-tool dynamometer is a dynamometer; it is used in measuring forces during the use of machine tools. The essential aspects of machine tools elucidated in the section are machining vibrations, machinist calculator, multimachine, ASME B5 etc. The section serves as a source to understand the major aspects related to machine tools.

Machine-Tool Dynamometer

A machine-tool dynamometer is a multi-component dynamometer that is used to measure forces during the use of the machine tool. Empirical calculations of these forces can be cross-checked and verified experimentally using these machine tool dynamometers.

With advances in technology, machine-tool dynamometers are increasingly used for the accurate measurement of forces and for optimizing the machining process. These multi-component forces are measured as an individual component force in each co-ordinate, depending on the coordinate system used. The forces during machining are dependent on depth of cut, feed rate, cutting speed, tool material and geometry, material of the work piece and other factors such as use of lubrication/cooling during machining.

Machining Vibrations

Machining vibrations, also called chatter, correspond to the relative movement between the workpiece and the cutting tool. The vibrations result in waves on the machined surface. This affects typical machining processes, such as turning, milling and drilling, and atypical machining processes, such as grinding.

A *chatter mark* is an irregular surface flaw left by a wheel that is out of true in grinding or regular mark left when turning a long piece on a lathe, due to machining vibrations.

As early as 1907, Frederick W. Taylor described machining vibrations as the most obscure and delicate of all the problems facing the machinist, an observation still true today, as shown in many publications on machining.

Mathematical models make it possible to simulate machining vibration quite accurately, but in practice it is always difficult to avoid vibrations and there are basic rules for the machinist:

- Rigidify the workpiece, the tool and the machine as much as possible
- Choose the tool that will excite vibrations as little as possible (modifying angles, dimensions, surface treatment, etc.)
- Choose exciting frequencies that best limit the vibrations of the machining system (spindle speed, number of teeth and relative positions, etc.)
- Choose tools that incorporate vibration-damping technology.

Industrial Context

Link between High-speed Machining and Vibrations

The use of high speed machining (HSM) has enabled an increase in productivity and the realization of workpieces that were impossible before, such as thin walled parts. Unfortunately, machine centers are less rigid because of the very high dynamic movements. In many applications, i.e. long tools, thin workpieces, the appearance of vibrations is the most limiting factor and compels the machinist to reduce cutting speeds well below the capacities of machines or tools.

Different Kinds of Problems and Their Sources

Vibration problems generally result in noise, bad surface quality and sometimes tool breakage. The main sources are of two types: forced vibrations and self-generated vibrations.

- Forced vibrations are mainly generated by interrupted cutting (inherent to milling), runout, or vibrations from outside the machine.
- Self generated vibrations are related to the fact that the actual chip thickness depends also on the relative position between tool and workpiece during the previous tooth passage. Thus increasing vibrations may appear up to levels which can seriously degrade the machined surface quality.

Laboratory Research

High-speed Strategies

Industrial and academic researchers have widely studied machining vibration. Specific strategies have been developed, especially for thin-walled work pieces, by alternating small machining passes in order to avoid static and dynamic flexion of the walls. The length of the cutting edge in contact with the workpiece is also often reduced in order to limit self-generated vibrations.

Modeling

The modeling of the cutting forces and vibrations, although not totally accurate, makes it possible to simulate problematic machining and reduce unwanted effects of vibration.

Stability Lobe Theory

Multiplication of the models based on stability lobe theory, which makes it possible to find the best spindle speed for machining, gives robust models for any kind of machining.

Time Domain Numerical Model

Time domain simulations compute workpiece and tool position on very small time scales without great sacrifice in accuracy of the instability process and of the surface modeled. These models need more computing resources than stability lobe models, but give greater freedom (cutting laws, runout, ploughing, finite element models). Time domain simulations are quite difficult to robustify, but a lot of work is being done in this direction in the research laboratories.

Paths

In addition to stability lobe theory, the use of variable tool pitch often gives good results, at a relatively low cost. These tools are increasingly proposed by tool manufacturers, although this is not really compatible with a reduction in the number of tools used. Other research leads are also promising, but often need major modifications to be practical in machining centers. Two kinds of software are very promising: Time domain simulations which give not yet reliable prediction but should progress, and vibration machining expert software, pragmatically based on knowledge and rules.

Industrial Methods Used to Limit Machining Vibrations

The Classic Approach

The usual method for setting up a machining process is still mainly based on historical technical knowhow and on trial and error method to determine the best parameters. According to the particular skills of a company, various parameters are studied in priority: depth of cut, tool path, workpiece set-up, geometrical definition of the tool,... When a vibration problem occurs, information is usually sought from the tool manufacturer or the CAM (Computer-aided manufacturing) software retailer, and they may give a better strategy for machining the workpiece. Sometimes, when vibration problems are too much of a financial prejudice, experts can be called upon to prescribe, after measurement and calculation, spindle speeds or tool modifications.

Limitations of The Available Methods

Compared to the industrial stakes, commercial solutions are rare. To analyse the problems and to propose solutions, only few experts propose their services. Computational software for stability lobes and measurement devices are proposed but, in spite of widespread publicity, they remain relatively rarely used. Lastly, vibration sensors are often integrated into machining centers but they are used mainly for wear diagnosis of the tools or the spindle. New Generation Tool Holders and especially the Hydraulic Expansion Tool Holders minimise the undesirable effects of vibration to a large extent. First of all, the precise control of total indicator reading to less than 3 micrometres helps reduce vibrations due to balanced load on cutting edges and the little vibration created thereon is absorbed largely by the oil inside the chambers of the Hydraulic Expansion Tool Holder.

Machinist Calculator

A machinist calculator is a hand-held calculator programmed with built-in formulas making it easy and quick for machinists to establish speeds, feeds and time without guesswork or conversion charts. Formulas may include revolutions per minute (RPM), surface feet per minute (SFM), inches per minute (IPM), feed per tooth (FPT). A cut time (CT) function takes the user, step-by-step, through a calculation to determine cycle time (execution time) for a given tool motion. Other features may include a metric-English conversion function, a stop watch/timer function and a standard math calculator.

This type of calculator is useful for machinists, programmers, inspectors, estimators, supervisors, and students.

Self-Replicating Machine

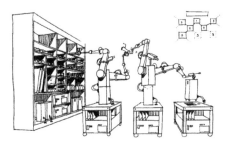

A simple form of machine self-replication

A self-replicating machine is a type of autonomous robot that is capable of reproducing itself autonomously using raw materials found in the environment, thus exhibiting self-replication in a way analogous to that found in nature. The concept of self-replicating machines has been advanced and examined by Homer Jacobsen, Edward F. Moore, Freeman Dyson, John von Neumann and in more recent times by K. Eric Drexler in his book on nanotechnology, *Engines of Creation* and by Robert Freitas and Ralph Merkle in their review *Kinematic Self-Replicating Machines* which provided the first comprehensive analysis of the entire replicator design space. The future development of such technology is an integral part of several plans involving the mining of moons and asteroid belts for ore and other materials, the creation of lunar factories, and even the construction of solar power satellites in space. The possibly misnamed von Neumann probe is one theoretical example of such a machine. Von Neumann also worked on what he called the universal constructor, a self-replicating machine that would operate in a cellular automata environment.

A self-replicating machine is an artificial self-replicating system that relies on conventional large-scale technology and automation. Certain idiosyncratic terms are occasionally found in the literature. For example, the term "clanking replicator" was once used by Drexler to distinguish macroscale replicating systems from the microscopic nanorobots or "assemblers" that nanotechnology may make possible, but the term is informal and is rarely used by others in popular or technical discussions. Replicators have also been called "von Neumann machines" after John von Neumann, who first rigorously studied the idea. However, the term "von Neumann machine" is less specific and also refers to a completely unrelated computer architecture that von Neumann proposed and so its use is discouraged where accuracy is important. Von Neumann himself used the term universal constructor to describe such self-replicating machines.

Historians of machine tools, even before the numerical control era, sometimes figuratively said that machine tools were a unique class of machines because they have the ability to "reproduce themselves" by copying all of their parts. Implicit in these discussions is that a human would direct the cutting processes (later planning and programming the machines), and would then be assembling the parts. The same is true for RepRaps, which are another class of machines sometimes mentioned in reference to such non-autonomous "self-replication". In contrast, machines that are *truly autonomously* self-replicating (like biological machines) are the main subject discussed here.

History

The general concept of artificial machines capable of producing copies of themselves dates back at least several hundred years. An early reference is an anecdote regarding the philosopher René Descartes, who suggested to Queen Christina of Sweden that the human body could be regarded as a machine; she responded by pointing to a clock and ordering "see to it that it reproduces offspring." Several other variations on this anecdotal response also exist. Samuel Butler proposed in his 1872 novel *Erewhon* that machines were already capable of reproducing themselves but it was man who made them do so, and added that *"machines which reproduce machinery do not reproduce machines after their own kind"*.

In 1802 William Paley formulated the first known teleological argument depicting machines producing other machines, suggesting that the question of who originally made a watch was rendered moot if it were demonstrated that the watch was able to manu-facture a copy of itself. Scientific study of self-reproducing machines was anticipated by John Bernal as early as 1929 and by mathematicians such as Stephen Kleene who began developing recursion theory in the 1930s. Much of this latter work was motivated by interest in information processing and algorithms rather than physical implementation of such a system, however.

Von Neumann's Kinematic Model

A detailed conceptual proposal for a physical non-biological self-replicating system was first put forward by mathematician John von Neumann in lectures delivered in 1948 and 1949, when he proposed a kinematic self-reproducing automaton model as a thought experiment. Von Neumann's concept of a physical self-replicating machine was dealt with only abstractly, with the hypothetical machine using a "sea" or stock-room of spare parts as its source of raw materials. The machine had a program stored on a memory tape that directed it to retrieve parts from this "sea" using a manipula-tor, assemble them into a duplicate of itself, and then copy the contents of its memory tape into the empty duplicate's. The machine was envisioned as consisting of as few as eight different types of components; four logic elements that send and receive stim-uli and four mechanical elements used to provide a structural skeleton and mobility. While qualitatively sound, von Neumann was evidently dissatisfied with this model of a self-replicating machine due to the difficulty of analyzing it with mathematical rigor. He went on to instead develop an even more abstract model self-replicator based on cellular automata. His original kinematic concept remained obscure until it was popu-larized in a 1955 issue of *Scientific American*.

Moore's Artificial Living Plants

In 1956 mathematician Edward F. Moore proposed the first known suggestion for a practical real-world self-replicating machine, also published in *Scientific American*. Moore's "artificial living plants" were proposed as machines able to use air, water and soil as sources of raw materials and to draw its energy from sunlight via a solar battery or a steam engine. He chose the seashore as an initial habitat for such machines, giving them easy access to the chemicals in seawater, and suggested that later generations of the machine could be designed to float freely on the ocean's surface as self-replicating factory barges or to be placed in barren desert terrain that was otherwise useless for in-dustrial purposes. The self-replicators would be "harvested" for their component parts, to be used by humanity in other non-replicating machines.

Dyson's Replicating Systems

The next major development of the concept of self-replicating machines was a series of thought experiments proposed by physicist Freeman Dyson in his 1970 Vanuxem Lecture. He proposed three large-scale applications of machine replicators. First was to send a self-replicating system to Saturn's moon Enceladus, which in addition to producing copies of itself would also be programmed to manufacture and launch solar sail-propelled cargo spacecraft. These spacecraft would carry blocks of Enceladean ice to Mars, where they would be used to terraform the planet. His second proposal was a solar-powered factory system designed for a terrestrial desert environment, and his third was an "industrial development kit" based on this replicator that could be sold to developing countries to provide them with as much industrial capacity as desired. When Dyson revised and reprinted his lecture in 1979 he added proposals for a modified version of Moore's seagoing artificial living plants that was designed to distill and store fresh water for human use and the "Astrochicken."

Advanced Automation for Space Missions

In 1980, inspired by a 1979 "New Directions Workshop" held at Wood's Hole, NASA conducted a joint summer study with ASEE entitled *Advanced Automation for Space Missions* to produce a detailed proposal for self-replicating factories to develop lunar resources without requiring additional launches or human workers on-site. The study was conducted at Santa Clara University and ran from June 23 to August 29, with the final report published in 1982. The proposed system would have been capable of exponentially increasing productive capacity and the design could be modified to build self-replicating probes to explore the galaxy.

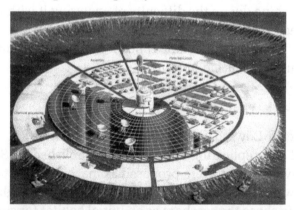

An artist's conception of a "self-growing" robotic lunar factory

The reference design included small computer-controlled electric carts running on rails inside the factory, mobile "paving machines" that used large parabolic mirrors to focus sunlight on lunar regolith to melt and sinter it into a hard surface suitable for building on, and robotic front-end loaders for strip mining. Raw lunar regolith would be refined by a variety of techniques, primarily hydrofluoric acid leaching. Large transports with a variety of manipulator arms and tools were proposed as the constructors that would put together new factories from parts and assemblies produced by its parent.

Power would be provided by a "canopy" of solar cells supported on pillars. The other machinery would be placed under the canopy.

A "casting robot" would use sculpting tools and templates to make plaster molds. Plaster was selected because the molds are easy to make, can make precise parts with good surface finishes, and the plaster can be easily recycled afterward using an oven to bake the water back out. The robot would then cast most of the parts either from nonconductive molten rock (basalt) or purified metals. A carbon dioxide laser cutting and welding system was also included.

A more speculative, more complex microchip fabricator was specified to produce the computer and electronic systems, but the designers also said that it might prove practical to ship the chips from Earth as if they were "vitamins."

A 2004 study supported by NASA's Institute for Advanced Concepts took this idea further. Some experts are beginning to consider self-replicating machines for asteroid mining.

Much of the design study was concerned with a simple, flexible chemical system for processing the ores, and the differences between the ratio of elements needed by the replicator, and the ratios available in lunar regolith. The element that most limited the growth rate was chlorine, needed to process regolith for aluminium. Chlorine is very rare in lunar regolith.

Lackner-wendt Auxon Replicators

In 1995, inspired by Dyson's 1970 suggestion of seeding uninhabited deserts on Earth with self-replicating machines for industrial development, Klaus Lackner and Christopher Wendt developed a more detailed outline for such a system. They proposed a colony of cooperating mobile robots 10–30 cm in size running on a grid of electrified ceramic tracks around stationary manufacturing equipment and fields of solar cells. Their proposal didn't include a complete analysis of the system's material requirements, but described a novel method for extracting the ten most common chemical elements found in raw desert topsoil (Na, Fe, Mg, Si, Ca, Ti, Al, C, O_2 and H_2) using a high-temperature carbothermic process. This proposal was popularized in Discover Magazine, featuring solar-powered desalination equipment used to irrigate the desert in which the system was based. They named their machines "Auxons", from the Greek word *auxein* which means "to grow."

Recent Work

Self-replicating Rapid Prototypers

Early experimentation with rapid prototyping in 1997-2000 was not expressly oriented toward reproducing rapid prototyping systems themselves, but rather extended simulated "evolutionary robotics" techniques into the physical world. Later developments in rapid prototyping have given the process the ability to produce a wide variety of electronic and mechanical components, making this a rapidly developing frontier in self-replicating system research.

RepRap 1.0 "Darwin" prototype

In 1998 Chris Phoenix informally outlined a design for a hydraulically powered replicator a few cubic feet in volume that used ultraviolet light to cure soft plastic feedstock and a fluidic logic control system, but didn't address most of the details of assembly procedures, error rates, or machining tolerances.

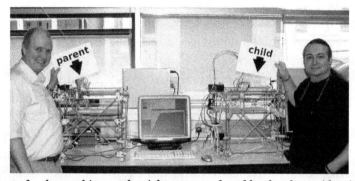

All of the plastic parts for the machine on the right were produced by the almost identical machine on the left. (Adrian Bowyer (left) and Vik Olliver(right) are members of the RepRap project.)

In 2005, Adrian Bowyer of the University of Bath started the RepRap Project to develop a rapid prototyping machine which would be able to manufacture some or most of its own components, making such machines cheap enough for people to buy and use in their homes. The project is releasing its designs and control programs under the GNU GPL. The RepRap approach uses fused deposition modeling to manufacture plastic components, possibly incorporating conductive pathways for circuitry. Other components, such as steel rods, nuts and bolts, motors and separate electronic components, would be supplied externally. In 2006 the project produced a basic functional prototype and in May 2008 the machine succeeded in producing all of the plastic parts required to make a 'child' machine.

Some researchers have proposed a microfactory of specialized machines that support recursion—nearly all of the parts of all of the machines in the factory can be manufactured by the factory.

Niac Studies on Self-Replicating Systems

In the spirit of the 1980 "Advanced Automation for Space Missions" study, the NASA Institute for Advanced Concepts began several studies of self-replicating system design in 2002 and 2003. Four phase I grants were awarded:

- Hod Lipson (Cornell University), "Autonomous Self-Extending Machines for Accelerating Space Exploration"

- Gregory Chirikjian (Johns Hopkins University), "Architecture for Unmanned Self-Replicating Lunar Factories"

- Paul Todd (Space Hardware Optimization Technology Inc.), "Robotic Lunar Ecopoiesis"

- Tihamer Toth-Fejel (General Dynamics), "Modeling Kinematic Cellular Automata: An Approach to Self-Replication" The study concluded that complexity of the development was equal to that of a Pentium 4, and promoted a design based on cellular automata.

Cornell University's Self-assembler

In 2005, a team of researchers at Cornell University, including Hod Lipson, implemented a self-assembling machine. The machine is composed of a tower of four articulated cubes, known as *molecubes*, which can revolve about a triagonal. This enables the tower to function as a robotic arm, collecting nearby molecubes and assembling them into a copy of itself. The arm is directed by a computer program, which is contained within each molecube, analogous to how each animal cell contains an entire copy of its DNA. However, the machine cannot manufacture individual molecubes, nor do they occur naturally, so its status as a self-replicator is debatable.

New York University Artificial DNA Tile Motifs

In 2011 a team of scientists at New York University created a structure called 'BTX' (bent triple helix) based around three double helix molecules, each made from a short strand of DNA. Treating each group of three double-helices as a code letter, they can (in principle) build up self-replicating structures that encode large quantities of information.

Self-replication of Magnetic Polymers

In 2001 Jarle Breivik at University of Oslo created a system of magnetic building blocks, which in response to temperature fluctuations, spontaneously form self-replicating polymers.

Self-replication of Neural Circuits

In 1968 Zellig Harris wrote that "the metalanguage is in the language," suggesting that self-replication is part of language. In 1977 Niklaus Wirth formalized this proposition by publishing a self-replicating deterministic context-free grammar. Adding to it probabilities, Bertrand du Castel published in 2015 a self-replicating stochastic grammar and presented a mapping of that grammar to neural networks, thereby presenting a model for a self-replicating neural circuit.

Partial Construction

Partial construction is the concept that the constructor creates a partially constructed (rather than fully formed) offspring, which is then left to complete its own construction.

The von Neumann model of self-replication envisages that the mother automaton should construct all portions of daughter automatons, without exception and prior to the initiation of such daughters. Partial construction alters the construction relationship between mother and daughter automatons, such that the mother constructs but a portion of the daughter, and upon initiating this portion of the daughter, thereafter retracts from imparting further influence upon the daughter. Instead, the daughter automaton is left to complete its own development. This is to say, means exist by which automatons may develop via the mechanism of a zygote.

Self-replicating Spacecraft

The idea of an automated spacecraft capable of constructing copies of itself was first proposed in scientific literature in 1974 by Michael A. Arbib, but the concept had appeared earlier in science fiction such as the 1967 novel *Berserker* by Fred Saberhagen or the 1950 novellette trilogy *The Voyage of the Space Beagle* by A. E. van Vogt. The first quantitative engineering analysis of a self-replicating spacecraft was published in 1980 by Robert Freitas, in which the non-replicating Project Daedalus design was modified to include all subsystems necessary for self-replication. The design's strategy was to use the probe to deliver a "seed" factory with a mass of about 443 tons to a distant site, have the seed factory replicate many copies of itself there to increase its total manufacturing capacity, and then use the resulting automated industrial complex to construct more probes with a single seed factory on board each.

Other References

- A number of patents have been granted for self-replicating machine concepts. The most directly relevant include U.S. Patent 4,734,856 "Autogeneric system"

Inventor: Davis; Dannie E. (Elmore, AL) (March 1988), U.S. Patent 5,659,477 "Self reproducing fundamental fabricating machines (F-Units)" Inventor: Collins; Charles M. (Burke, VA) (August 1997), U.S. Patent 5,764,518 " Self reproducing fundamental fabricating machine system" Inventor: Collins; Charles M. (Burke, VA)(June 1998); Collins' PCT: and U.S. Patent 6,510,359 "Method and system for self-replicating manufacturing stations" Inventors: Merkle; Ralph C. (Sunnyvale, CA), Parker; Eric G. (Wylie, TX), Skidmore; George D. (Plano, TX) (January 2003).

- Macroscopic replicators are mentioned briefly in the fourth chapter of K. Eric Drexler's 1986 book *Engines of Creation*.

- In 1995, Nick Szabo proposed a challenge to build a macroscale replicator from Lego robot kits and similar basic parts. Szabo wrote that this approach was easier than previous proposals for macroscale replicators, but successfully predicted that even this method would not lead to a macroscale replicator within ten years.

- In 2004, Robert Freitas and Ralph Merkle published the first comprehensive review of the field of self-replication (from which much of the material in this article is derived, with permission of the authors), in their book *Kinematic Self-Replicating Machines*, which includes 3000+ literature references. This book included a new molecular assembler design, a primer on the mathematics of replication, and the first comprehensive analysis of the entire replicator design space.

In Fiction

In Literature

Many types of fictional self-replicating machines have been featured in literature, and particularly in science fiction.

Fictional self-replicating machines in literature			
Year	**Work**	**Author**	**Notes**
1943	"M33 in Andromeda"	A. E. van Vogt	A. E. van Vogt used the idea as a plot device in his story "M33 in Andromeda" (1943) which was later combined with the three other *Space Beagle* short stories to become the novel, The Voyage of the Space Beagle. The story describes the creation of self-replicating weapons factories designed to destroy the Anabis, a galaxy-spanning malevolent life form bent on destruction of the human race.

Fictional self-replicating machines in literature			
Year	**Work**	**Author**	**Notes**
1953	"Second Variety"	Philip K. Dick	In the short story a nuclear war between the Soviet Union and the West has reduced much of the world to a barren wasteland. The war continues, however, among the scattered remains of humanity. The Western forces have developed "claws", which are autonomous self-replicating robots to fight on their side. It is one of Dick's many stories in which nuclear war has rendered the Earth's surface uninhabitable. The story was adapted into the movie *Screamers* in 1995.
1955	"Autofac"	Philip K. Dick	An early treatment was the short story *Autofac* by Philip K. Dick, published in 1955. Dick also touched on this theme in his earlier 1953 short story "Second Variety". Another example can be found in the 1962 short story "Epilogue" by Poul Anderson, in which self-replicating factory barges were proposed that used minerals extracted from ocean water as raw materials.
1955	"The Necessary Thing"	Robert Sheckley	In the short story the Universal Replicator is unwittingly tricked into replicating itself.
1958	"Crabs on the Island"	Anatoly Dneprov	In his short story "Crabs on the Island" (1958) Anatoly Dneprov speculated on the idea that since the replication process is never 100% accurate, leading to slight differences in the descendants, over several generations of replication the machines would be subjected to evolution similar to that of living organisms. In the story, a machine is designed, the sole purpose of which is to find metal to produce copies of itself, intended to be used as a weapon against an enemy's war machines. The machines are released on a deserted island, the idea being that once the available metal is all used and they start fighting each other, natural selection will enhance their design. However, the evolution has stopped by itself when the last descendant, an enormously large crab, was created, being unable to reproduce itself due to lack of energy and materials.

Fictional self-replicating machines in literature			
Year	**Work**	**Author**	**Notes**
1963-2005	*Berserker* series	Fred Saberhagen	The *Berserker* series is a series of space opera science fiction short stories and novels, in which robotic self-replicating machines (The berserkers) strive to destroy all life.
1964	*The Invincible*	Stanisław Lem	Stanisław Lem has also studied the same idea in his novel, in which the crew of a spacecraft landing on a distant planet finds a non-biological life-form, which is the product of long, possibly of millions of years of, mechanical evolution (necro-evolution). This phenomenon is also key to the aforementioned Anderson story.
1968	*The Reproductive System*	John Sladek	John Sladek used the concept to humorous ends in his first novel *The Reproductive System* (1968, also titled *Mechasm* in some markets), where a U.S. military research project goes out of control.
1970	"The Scarred Man"	Gregory Benford	Long before the existence of the Internet, author Greg Benford was inspired by his work on ARPANet in the late 1960s to write this first account of a self-replicating program - a computer virus. His con men program a computer to randomly dial phone numbers until it hits a telephone modem that is answered by another computer. It then programs the answering computer to begin dialing random numbers in search of yet another computer, while also programming a small delay on each computer's processing time. The virus spreads exponentially through susceptible computers, like a biological infection, and the creators profit by "fixing" the slowed computers. (Story text on author's website.)
1975	*The Shockwave Rider*	John Brunner	An early example of a fictional account of a computer virus or worm.
1977	*The Adolescence of P-1*	Thomas J. Ryan	Another early fictional account of a computer virus or worm.

	Fictional self-replicating machines in literature		
Year	**Work**	**Author**	**Notes**
1977-1999	*Galactic Center Saga* series	Gregory Benford	The series details a galactic war between mechanical and biological life. In it an antagonist berserker machine race is encountered by Earth, first as a probe in *In the Ocean of Night*, and then in an attack in *Across the Sea of Suns*. The berserker machines do not seek to completely eradicate a race if merely throwing it into a primitive low technological state will do as they did to the EMs encountered in *Across the Sea of Suns*.
1982	*2010: Odyssey Two*	Arthur C. Clarke	The novel is the sequel to the 1968 novel *2001: A Space Odyssey*, but continues the story of Stanley Kubrick's film adaptation with the same title rather than Clarke's original novel. Set in the year 2010, the plot centers on a joint Soviet-American mission aboard the Soviet spacecraft *Leonov*. Its crew flees Jupiter as a mysterious dark spot appears on Jupiter and begins to grow. HAL's telescope observations reveal that the "Great Black Spot" is, in fact, a vast population of monoliths, increasing at an exponential rate, which appear to be eating the planet. By acting as self-replicating 'von Neumann' machines, these monoliths increase Jupiter's density until the planet achieves nuclear fusion, becoming a small star.
1983	*Code of the Lifemaker*	James P. Hogan	NASA's Advanced Automation for Space Missions study directly inspired the science fiction novel.
1985	*The Third Millennium: A History of the World AD 2000-3000*	Brian Stableford David Langford	In the book—a fictional historical account, from the perspective of the year 3000, giving a future history of humanity and its technological and sociological developments—humanity sends cycle-limited Von Neumann probes out to the nearest stars to do open-ended exploration and to announce humanity's existence to whoever might encounter them.

Fictional self-replicating machines in literature			
Year	**Work**	**Author**	**Notes**
1986	*The Songs of Distant Earth*	Arthur C. Clarke	In the novel humanity on a future Earth facing imminent destruction creates automated seedships that act as fire and forget lifeboats aimed at distant, habitable worlds. Upon landing, the ship begins to create new humans from stored genetic information, and an onboard computer system raises and trains the first few generations of new inhabitants. The massive ships are then broken down and used as building materials by their "children".
1986	"Lungfish"	David Brin	In the short story collection, *The River of Time*, the short story "Lungfish" prominently features von Neumann probes. Not only does he explore the concept of the probes themselves, but indirectly explores the ideas of competition between different designs of probes, evolution of von Neumann probes in the face of such competition, and the development of a type of ecology between von Neumann probes. One of the vessels mentioned is clearly a Seeder type.
1987	*The Forge of God*	Greg Bear	The *Killers*, a civilization of self-replicating machines designed to destroy any potential threat to their (possibly long-dead) creators.
1990	*The World at the End of Time*	Frederik Pohl	
1992	*Cold as Ice*	Charles Sheffield	In the novel there is a segment where the author (a physicist) describes von Neumann machines harvesting sulfur, nitrogen, phosphorus, helium-4, and various metals from the atmosphere of Jupiter.

Fictional self-replicating machines in literature			
Year	Work	Author	Notes
1993	*Assemblers of Infinity*	Kevin J. Anderson and Doug Beason	This novel describes self-replicating robots that are programmed not to harm biospheres but instead use materials on the moon for an alien civilization to reproduce and colonize the moon. While this is happening a human scientist on Earth reverse engineers the dormant nanomachines found on Earth (since Earth is a biosphere they don't harm the environment) to make medical nano-machines and is successful at first when he revives a medically dead scientist, but accidentally removes the safety measure, creating a grey goo scenario that he stops at the cost of his life when he activates a high powered x-ray machine built as a safety guard.
1993	*Anvil of Stars*	Greg Bear	The novel is the sequel to *The Forge of God* and explores the reaction other civilizations have to the creation and release of berserkers.
1995	*The Ganymede Club*	Charles Sheffield	A mystery and a thriller, the story unravels in the same universe that Sheffield imagined in *Cold as Ice*. In it humans have colonized the solar system with the help of self-replicating machines called Von Neumanns.
1995	*The Diamond Age*	Neal Stephenson	The novel depicts a near-future Earth society wherein nanotechnology, including self-replicators, both exist and influence daily life greatly.
1996	*Excession*	Iain Banks	In the novel hegemonising swarms are described as a form of Outside Context Problem. An example of an "Aggressive Hegemonising Swarm Object" is given as an uncontrolled self-replicating probe with the goal of turning all matter into copies of itself. After causing great damage, they are somehow transformed using unspecified techniques by the Zetetic Elench and become "Evangelical Hegemonising Swarm Objects". Such swarms (referred to as "smatter") reappear in the later novels *Surface Detail* (which features scenes of space combat against the swarms) and *The Hydrogen Sonata*.

Fictional self-replicating machines in literature			
Year	**Work**	**Author**	**Notes**
1998	*Moonseed*	Stephen Baxter	In the novel Earth faces danger from a self-replicating nanobot swarm after a rock is returned from the Apollo 18 mission. The rock contains a mysterious substance called "moonseed" (a form of grey goo, whether nanobots, an alien virus or something else) that starts to change all inorganic matter on Earth into more moonseed.
1998	*Bloom*	Wil McCarthy	*Bloom* is set in the year 2106, in a world where self-replicating nanomachines called "Mycora" have consumed Earth and other planets of the inner solar system, forcing humankind to eke out a bleak living in the asteroids and Galilean moons.
1998	*Destiny's Road*	Larry Niven	In the novel von Neumann machines are scattered throughout the human colony world Destiny and its moon Quicksilver in order to build and maintain technology and to make up for the lack of the resident humans' technical knowledge; the Von Neumann machines primarily construct a stretchable fabric cloth capable of acting as a solar collector which serves as the humans' primary energy source. The Von Neumann machines also engage in ecological maintenance and other exploratory work.
2000	*Manifold: Space*	Stephen Baxter	The novel starts with the discovery of alien self-replicating machines active within the Solar system.

Fictional self-replicating machines in literature			
Year	**Work**	**Author**	**Notes**
2000–present	*Revelation Space* series	Alastair Reynolds	In the series Inhibitors are self-replicating machines whose purpose is to inhibit the development of intelligent star-faring cultures. They are dormant for extreme periods of time until they detect the presence of a space-faring culture and proceed to exterminate it even to the point of sterilizing entire planets. They are very difficult to destroy as they seem to have faced any type of weapon ever devised and only need a short time to 'remember' the necessary counter-measures. Also "Greenfly" terraforming machines are another form of berserker machines. For unknown reasons, but probably an error in their programming, they destroy planets and turn them into trillions of domes filled with vegetation – after all, their purpose is to produce a habitable environment for humans, however in doing so they inadvertently decimate the human race. By 10.000, they have wiped out most of the Galaxy.
2002	*Evolution*	Stephen Baxter	The novel follows 565 million years of human evolution, from shrewlike mammals 65 million years in the past to the ultimate fate of humanity (and its descendants, both biological and non-biological) 500 million years in the future. At one point, hominids become sapient, and go on to develop technology, including an evolving universal constructor machine that goes to Mars and multiplies, and in an act of global ecophagy consumes Mars by converting the planet into a mass of machinery that leaves the Solar system in search of new planets to assimilate.
2002	*Prey*	Michael Crichton	In the novel nanobots were blown into the desert from an isolated laboratory, evolving and eventually forming autonomous swarms. These swarms appear to be clouds of solar-powered and self-sufficient, reproducing and evolving rapidly. The swarms exhibit predatory behavior, attacking and killing animals in the wild and later going as far as to forming symbiotic relationships with humans and even mimicking them.

Fictional self-replicating machines in literature			
Year	**Work**	**Author**	**Notes**
2002	*Lost in a Good Book*	Jasper Fforde	The novel features an alternative *pink goo* end of the world scenario, where a nanotechnology 'Dream Topping making machine' turns all matter on earth into a pink dessert similar to Angel Delight. The Dream Topping is taken back in time to the beginning of earth, where it supplies the organic nutrients needed to create life.
2003	*Ilium*	Dan Simmons	The first part of the *Ilium/Olympos* cycle, concerning the re-creation of the events in the *Iliad* on an alternate Earth and Mars. These events are set in motion by beings who have taken on the roles of the Greek gods. In the cycle the *voynix* are biomechanical, self-replicating, programmable robots. They originated in an alternate universe, and were brought into the Ilium universe before 3000 A.D.
2003	*Singularity Sky*	Charles Stross	The *Festival*, a civilisation of uploaded minds with strange designs on humanity
2004	*Recursion*	Tony Ballantyne	Herb, a young entrepreneur, returns to the isolated planet on which he has illegally been trying to build a city–and finds it destroyed by a swarming nightmare of self-replicating machinery.
2005	*Spin*	Robert Charles Wilson	In the novel self-replicating artificial life, shot into space to build a huge sentient network in the outer reaches of the Solar System and gather information about the alien "Hypotheticals". It encounters not just other von Neumann machines, but a pre-existing and galaxy-spanning ecology of them. Apparently this vast network of sentient artificial life is responsible for the "Spin" – the placement of an opaque black membrane around the entire Earth.
2005	*Olympos*	Dan Simmons	The sequel to *Ilium* and final part of the *Ilium/Olympos* series.

Fictional self-replicating machines in literature			
Year	**Work**	**Author**	**Notes**
2007	*Von Neumann's War*	John Ringo Travis S. Taylor	In the novel published by Baen Books in 2007 von Neumann probes arrive in the solar system, moving in from the outer planets, and converting all metals into gigantic structures. Eventually they arrive on Earth, wiping out much of the population before being beaten back when humanity reverse engineers some of the probes.
2007	*Postsingular*	Rudy Rucker	In Postsingular, nanobots devour the Earth and copy everybody they eat into a simulation... luckily, one of the machine's developers also created a backdoor, and is able to reverse the situation, restoring everybody. Soon after, another set of tiny self-replicating machines are released, which don't devour, merely reproduce until they cover every inch of the Earth, sharing information with each other and the people they're on. They connect humanity like they've never been connected before so that one can watch anyone else by experiencing what the "orphids" on that person's body are experiencing.
2010	*Surface Detail*	Iain Banks	The novel depicts self-replicating machines as a universe-threatening infection.
2011	*Lord of All Things*	Andreas Eschbach	In the novel (original title "Herr aller Dinge") an ancient nano machine complex is discovered buried in a glacier off the coast of Russia. When it comes in contact with materials it needs to fulfill its mission, it creates a launch facility and launches a space craft. It is later revealed that the nano machines were created by a pre-historic human race with the intention of destroying other interstellar civilizations (for an unknown reason). It is purposed that the reason there is no evidence of the race is because of the nano-machines themselves and their ability to manipulate matter at an atomic level. It is even suggested that viruses could be ancient nano machines that have evolved over time.
2012	*The Hydrogen Sonata*	Iain Banks	

Fictional self-replicating machines in literature			
Year	**Work**	**Author**	**Notes**
2012– pres- ent	*The Machine Dynasty* series	Madeline Ashby	In the novels the protagonists are von Neumann machines, self-replicating humanoid robots. The original proposal for the self-replicating humanoid robots came from a religious End Times group who wanted to leave a body of helpers behind for the millions of unsaved after the rapture.
2014	*Creations*	William Mitchell	In the novel biological engineer Max Lowrie gets a job offer of a lifetime that's supposed to pave the way for humanity's future: self-replicating machines that can mine materials from the harshest environments at no cost, opening up as yet unheard of resources in the sea, on land, and ultimately on the Moon.

In Film

Many types of self-replicating machines have been featured in the movies.

- The movie *Screamers*, based on Philip K. Dick's short story "Second Variety", features a group of robot weapons created by mankind to act as Von Neumann devices / berserkers. The original robots are subterranean buzzsaws that make a screaming sound as they approach a potential victim beneath the soil. These machines are self-replicating and, as is found out through the course of the movie, they are quite intelligent and have managed to "evolve" into newer, more dangerous forms, most notably human forms which the real humans in the movie cannot tell apart from other real humans except by trial and error.

- *The Terminator* is a 1984 science fiction/action film directed and co-written by James Cameron which describes a war between mankind and self replicating machines led by a central artificial intelligence known as Skynet. Machine civilizations are a recurring theme in fiction.

- In the 2008 science fiction film *The Day the Earth Stood Still* follows Klaatu, an alien sent to try to change human behavior or eradicate them from Earth via grey goo due to humankind's environmental damage to the planet.

- The *Star Wars* expanded universe features the World Devastators, large ships designed and built by the Galactic Empire that tear apart planets to use its materials to build other ships or even upgrade or replicate themselves.

- The Matrix series features self-replicating nanobots that were deployed by the United Nations to block the rebelling machine population from their energy source.

On Television

The concept is also widely utilised in science fiction television.

- The TV series *Lexx* featured an army of self replicating robots known as Mantrid drones.

- The Replicators are a horde of self-replicating machines that appear frequently in *Stargate SG-1*. They once were a vicious race of insect-like robots that were originally created by an android named Reese to serve as toys. They grew beyond her control and began evolving, eventually spreading throughout at least two galaxies. In addition to ordinary autonomous evolution they were able to analyze and incorporate new technologies they encountered into themselves, ultimately making them one of the most advanced "races" known. During the course of the series, the replicators assume a human form and pose a huge threat to the galaxy. A more sophisticated version of the human form Replicators, who call themselves Asurans also appear in the spin-off series *Stargate Atlantis*.

    o In the *Stargate SG-1* episode "Scorched Earth", a species of newly relocated humanoids face extinction via an automated terraforming colony seeder ship controlled by an artificial intelligence.

- In *Stargate Atlantis*, a second race of replicators created by the Ancients were encountered in the Pegasus Galaxy. They were created as a means to defeat the Wraith. The Ancients attempted to destroy them after they began showing sings of sentience and requested that their drive to kill the wraith be removed. This failed, and an unspecified length of time after the Ancients retreated to the Milky Way Galaxy, the replicators nearly succeeded in destroying the Wraith. The Wraith were able to hack into the replicators and deactivate the extermination drive, at which point they retreated to their home world and were not heard from again until encountered by the Atlantis Expedition. After the Atlantis Expedition reactivated this dormant directive, the replicators embarked on a plan to kill the Wraith by removing their food source, i.e. all humans in the Pegasus Galaxy.

- In the *Stargate Universe* the human adventurers live on a ship called *Destiny*. Its mission was to connect a network of Stargates, placed by preceding seeder ships, on planets capable of supporting life to allow instantaneous travel between them.

    o In *Stargate Universe* Season 2, a galaxy billions of light years distant from the Milky Way is infested with drone ships that are programmed to annihilate intelligent life and advanced technology. The drone ships attack other space ships (including Destiny) as well as humans on planetary surfaces, but don't bother destroying primitive technology such as buildings unless they are harboring intelligent life or advanced technology.

- *Star Trek's* Borg – a self-replicating bio-mechanical race that is dedicated to the task of achieving perfection through the assimilation of useful technology and lifeforms. Their ships are massive mechanical cubes (a close step from the Berzerker's massive mechanical Spheres).

- The episodes "A Clockwork Origin" and "Benderama" of the animated science fiction comedy sitcom *Futurama*.

- The episode "Walkabout" of *Gargoyles* (season 2, episode 33) is about grey goo.

- The *Babylon 5* episode "Infection" showed a smaller scale berserker in the form of the Icarran War Machine. After being created with the goal of defeating an unspecified enemy faction, the War Machines proceeded to exterminate all life on the planet Icarra VII because they had been programmed with standards for what constituted a 'Pure Icaran' based on religious teachings, which no actual Icaran could satisfy. Because the Icaran were pre-starflight, the War Machines became dormant after completing their task rather than spreading. One unit was reactivated on-board Babylon 5 after being smuggled past quarantine by an unscrupulous archaeologist, but after being confronted with how they had rendered Icara VII a dead world, the simulated personality of the War Machine committed suicide.

- In the *Justice League Unlimited* episode "Dark Heart", an alien weapon based on the idea lands on Earth.

- In *Steven Universe*, Gems are a race of artificial intelligences composed of gemstones projecting light-construct bodies. These are created by bacterio-phage-like Injector engines that drill into a planet's crust and infuse specific gems with the local biota's life energy, animating it; they do not reproduce naturally, and several similarities to computers have been noticed.

In Video Games

- *Grey Goo* is a science fiction real-time strategy video game that features a playable faction based on the grey goo scenario.

- *Conway's Game of Life*

- *Tasty Planet*, a game released in 2006 by Dingo Games centers around a gray goo eating the universe, starting at the atomic level and progressing to the cosmic level. In the game the player controls a gray goo and eats many objects, such as bacteria, mice, cars, people, Earth, galaxies, and eventually the universe. In the end, the grey goo over-fills, explodes, and starts the universe all over again.

- *Hostile Waters: Antaeus Rising*

- The Reapers in the video game series *Mass Effect* are also self-replicating probes bent on destroying any advanced civilization encountered in the galaxy. They lie dormant in the vast spaces between the galaxies and follow a cycle of extermination. In *Mass Effect 2* it is shown that they assimilate any advanced species.

- In the computer game *Star Control II*, the Slylandro Probe is an out-of-control self-replicating probe that attacks starships of other races. They were not originally intended to be a berserker probe; they sought out intelligent life for peaceful contact, but due to a programming error, they would immediately switch to "resource extraction" mode and attempt to dismantle the target ship for raw materials. While the plot claims that the probes reproduce "at a geometric rate", the game itself caps the frequency of encountering these probes. It is possible to deal with the menace in a side-quest, but this is not necessary to complete the game, as the probes only appear one at a time, and the player's ship will eventually be fast and powerful enough to outrun them or destroy them for resources – although the probes will eventually dominate the entire game universe.

- In the *Homeworld: Cataclysm* video game, a bio-mechanical virus called *Beast* has the ability to alter organic and mechanic material to suit its needs, and the ships infected become self-replicating hubs for the virus.

- In the computer game *Sword of the Stars*, the player may randomly encounter "Von Neumann". A Von Neumann mothership appears along with smaller Von Neumann probes, which attack and consume the player's ships. The probes then return to the mothership, returning the consumed material. If probes are destroyed, the mothership will create new ones. If all the player's ships are destroyed, the Von Neumann probes will reduce the planets resource levels before leaving. The mothership is a larger version of the probes. In the 2008 expansion *A Murder of Crows*, Kerberos Productions also introduces the VN Berserker, a combat orientated ship, which attacks player planets and ships in retaliation to violence against VN Motherships. If the player destroys the Berserker things will escalate and a System Destroyer will attack.

- In the *X* video game series, the Xenon are a malevolent race of artificially intelligent machines descended from terraforming ships sent out by humans to prepare worlds for eventual colonization; the result caused by a bugged software update. They are continual antagonists in the X-Universe.

- In PC role-playing game *Space Rangers* and its sequel *Space Rangers 2: Dominators*, a league of 5 nations battles three different types of Berserker robots. One that focuses on invading planets, another that battles normal space and third that lives in hyperspace.

- In the *Star Wolves* video game series, Berserkers are a self-replicating machine menace that threatens the known universe for purposes of destruction and/or assimilation of humanity.

- In the *Metroid* video game series, The massive Leviathans are probes routinely sent out from the planet Phaaze to infect other planets with Phazon radiation and eventually turn these planets into clones of Phaaze, where the self-replication process can continue.

- In the second Deus Ex game, *Deus Ex: Invisible War*, a videogame features a self-replicating nanomachines bomb in the CGI introduction. A terrorist attack on Boston erased the town and is the beginning of the plot.

Other

- The idea dates back at least as far as Karel Capek's 1920 play *R.U.R. (Rossum's Universal Robots)*. A fundamental obstacle of self-replicating machines, how to repair the repair systems, was the critical failure in the automated society described in *The Machine Stops*.

- In the comic *Transmetropolitan* a character mentions "Von Neumann rectal infestations" which are apparently caused by "Shit-ticks that build more shit-ticks that build more shit-ticks".

- In the anime *Vandread*, harvester ships attack vessels from both male- and female-dominated factions and harvest hull, reactors, and computer components to make more of themselves. To this end, Harvester ships are built around mobile factories. Earth-born humans also view the inhabitants of the various colonies to be little more than spare parts.

- In the role-playing game *Eclipse Phase*, an ETI probe is believed to have infected the TITAN computer systems with the Exsurgent virus to cause them to go berserk and wage war on humanity. This would make ETI probes a form of berserker, albeit one that uses pre-existing computer systems as its key weapons.

- *Storm*, the trilogy of albums which conclude the comic book series *Storm* by Don Lawrence (starting with *Chronicles of Pandarve 11: The Von Neumann machine*) is based on self-replicating conscious machines containing the sum of all human knowledge employed to rebuild human society throughout the universe in case of disaster on Earth. The probe malfunctions and although new probes are built, they do not separate from the motherprobe which eventually results in a cluster of malfunctioning probes so big that it can absorb entire moons.

- Denial-of-service attacks in the virtual world Second Life which work by continually replicating objects until the server crashes are referred to as *gray goo* attacks. This is a reference to the self-replicating aspects of gray goo. It is one example of the widespread convention of drawing analogies between certain Second Life concepts and the theories of radical nanotechnology.

- In the manga *Battle Angel Alita: Last Order*, the surface of Mercury is covered in rogue nanomachines from a Gray Goo event and subsequently spawns a being of dubious morphology known as Anomaly.

- *Baldr Sky*, a Japanese visual novel features self-replicating nanomachines under the name "assemblers". Their inherent threat and a near-catastrophic grey goo event is central to its plot.

Prospects for Implementation

As the use of industrial automation has expanded over time, some factories have begun to approach a semblance of self-sufficiency that is suggestive of self-replicating machines. However, such factories are unlikely to achieve "full closure" until the cost and flexibility of automated machinery comes close to that of human labour and the manufacture of spare parts and other components locally becomes more economical than transporting them from elsewhere. As Samuel Butler has pointed out in Erewhon, replication of partially closed universal machine tool factories is already possible. Since safety is a primary goal of all legislative consideration of regulation of such development, future development efforts may be limited to systems which lack either control, matter, or energy closure. Fully capable machine replicators are most useful for developing resources in dangerous environments which are not easily reached by existing transportation systems (such as outer space).

An artificial replicator can be considered to be a form of artificial life. Depending on its design, it might be subject to evolution over an extended period of time. However, with robust error correction, and the possibility of external intervention, the common science fiction scenario of robotic life run amok will remain extremely unlikely for the foreseeable future.

Multimachine

The multimachine is an all-purpose open source machine tool that can be built inexpensively by a semi-skilled mechanic with common hand tools, from discarded car and truck parts, using only commonly available hand tools and no electricity. Its size can range from being small enough to fit in a closet to one hundred times that size. The multimachine can accurately perform all the functions of an entire machine shop by itself.

MultiMachine

The multimachine was first developed as a personal project by Pat Delaney, then grew into an open source project organized via a Yahoo! group. The 2,600 member support group that has grown up around its creation is made up of engineers, machinists, and experimenters who have proven that the machine works. As an open-source machine tool that can be built cheaply on-site, the Multimachine could have many uses in developing countries. The multimachine group is currently focused on the humanitarian aspects of the multimachine, and on promulgating the concept of the multimachine as a means to create jobs and economic growth in developing countries.

The multimachine first became known to a wider audience as the result of the 2006 Open Source Gift Guide article on the *Make* magazine website, in which the multimachine was mentioned under the caption "Multimachine - Open Source machine tool".

Uses

As a general-purpose machine tool that includes the functions of a milling machine, drill press, and lathe, the multimachine can be used for many projects important for humanitarian and economic development in developing countries:

- Agriculture: Building and repairing irrigation pumps and farm implements

- Water supplies: Making and repairing water pumps and water-well drilling rigs.

- Food supplies: Building steel-rolling-and-bending machines for making fuel efficient cook stoves and other cooking equipment

- Transportation: Anything from making cart axles to rebuilding vehicle clutch, brake, and other parts.

- Education: Building simple pipe-and-bar-bending machines to make school furniture, providing "hands on" training on student-built multimachines that they take with them when they leave school.

- Job creation: A group of specialized but easily built multimachines can be combined to form a small, very low cost, metal working factory which could also serve as a trade school. Students could be taught a single skill on a specialized machine and be paid as a worker while learning other skills that they could take elsewhere.

Accuracy

The design goals of the multimachine were to create an easily built machine tool, made from "junk," that is nonetheless all-purpose and accurate enough for production work. It has been reported to be able to make cuts within a tenth (one ten-thousandth of an inch), which means that in at least some setups it can equal commercial machine tool accuracy.

In almost every kind of machining operation, either the work piece or the cutting tool turns. If enough flexibility is built into the parts of a machine tool involved in these functions, the resulting machine can do almost every kind of machining operation that will physically fit on it. The multimachine starts with the concept of 3-in-1 machine tools—basically a combination of metal lathe, mill and drill press—but adds many other functions. It can be a 10-in-1 (or even more) machine tool.

Construction

At a high-level, the multimachine is built using vehicle engine blocks combined in a LEGO-like fashion. It utilizes the cylinder bores and engine deck to provide accurate surfaces. Since cylinder bores are bored exactly parallel to each other and at exact right angles to the cylinder head surface, multimachine accuracy begins at the factory where the engine block was built. In the most common version of the multimachine, one that has a roller bearing spindle, this precision is maintained during construction with simple cylinder re-boring of the #3 cylinder to the size of the roller bearing outside diameter (OD) and re-boring the #1 cylinder to fit the overarm OD. These cylinder-boring operations can be done in almost any engine shop and at low cost. An engine machine shop provides the most inexpensive and accurate machine work commonly done anywhere and guarantees that the spindle and overarm will be perfectly aligned and at an exact right angle to the face (head surface) of the main engine block that serves as the base of the machine. Use a piece of pipe made to fit the inner diameter of the bearings as the spindle. A three-bearing spindle is used because the "main" spindle bearings just "float" in the cylinder bore so that the third bearing is needed to "locate" the spindle, act as a thrust bearing, and support the heavy pulley. The multimachine uses a unique way of clamping the engine blocks together that is easily built, easily adjusted, and very accurate. The multimachine makes use of a concrete and steel construction technique that was heavily used in industry during the First World War and resurrected for this project.

ASME B5

ASME B5 refers to a technical committee of the American Society of Mechanical Engineers and the standard they maintain which deals with machine tools.

Asme (American Society of Mechanical Engineers) B5 Machine Tools - Components, Elements, Performance, and Equipment

As a Standards Development Organization, ASME continues to develop and maintains nearly 600 codes and standards in a wide range of disciplines including pressure technology, nuclear plants, elevators / escalators, construction, engineering design, and performance testing. Machine Tool standards are developed and maintained by ASME B5 Committee, which operates under ASME's Board on Standardization and Testing. The B5 Standards Committee currently meets once a year in various locations throughout the United States. The meeting is generally held in November and is open to the public. The B5 Technical Committees usually meet in conjunction with the B5 Standards Committee. Some B5 Technical Committees also meet separately in different locations throughout the year. The B5 Standards Committee and its Technical Committees are composed of experts in the field of machine tools. Members of the B5 Standards Committee are classified in the following interest classes: Producer/Manufacturer, Regulatory, Services, General Interest, and User. The B5 Committee works on writing new ASME American National Standards, and revising current ASME B5 and B94 standards. The B5 Standards Committee operates under procedures of the American National Standards Institute (ANSI).

Turn of the Century Machine Shop

ASME B5 History

The celebration of the 100th meeting in 2022 will mark the 100th consecutive year the ASME B5 committee has met. The following historical information marks only the beginning stages of copious collaborative work and continuous publication, review, and revision of ASME B5 Standards.

On May 13, 1914 the ASME Committee on Meetings, Subcommittee on Machine Shop Practice began discussion on standardization of machine shop practices. It wasn't until September 1922 under the procedure of American Standards Association, the B5 committee was organized as a committee dedicated to machine tools. B5 was sponsored by the National Machine Tool Builders' Association, the Society of Automotive Engineers, Metal Cutting Tool Institute, and The American Society of Mechanical Engineers.

On December 2, 1937, the standard for Adjustable Adapters for Multiple Spindle Drilling Heads this standard was approved by the American Standards Association and designated as American Standard (ASA B5.11-1937).

Work on the standardization of T-slots started in 1924 and a tentative standard was published in 1927. The first official American Standard for T-slots came in 1941. B5 Technical Committee No. 11 was organized in New York on December 4, 1928 and B5 Technical Committee No. 4 on Spindle Noses was organized on December 5, 1928. These two committees worked in close cooperation with each other and with manufacturers and users of engine lathes, turret lathes and automatic lathes in developing standards for spindle noses and chucks.

B5 Technical Committee No. 3 on the Standardization of Machine Tapers was appointed in August, 1926,and held its organization meeting in September, 1926, in New Haven, Conn. Three American tapers then in use, the Brown & Sharpe(1860), Morse (1862), and Jarno (1889), and the taper series adopted by William Sellers & Co.(1862) were combined into a compromise standard series which contained twenty-two (22) self-holding taper sizes.

The first edition of the Spindle Noses and Tool Shanks for Milling Machines standard, known as B5.18-1943, resulted from intensive effort dating back to 1926 by a special group of milling machine manufacturers.

The effort to establish an American standard for Ball Screws began in July 1971 out of a need to obtain a consensus opinion relative to proposals for standardization of ball screw assemblies within the ISO/TC39 sub-committee Working Group 7. Out of this, the subcommittee TC43 was organized with members representing manufacturers, users of ball screws, and others of general interest. TC43 produced ANSI B5.48, which was approved as an American National Standard and published in 1977.

ASME B5 Charter

The charter of the ASME B5 Machine Tool Standards committee is "The standardization of machine tools, cutting tools and of the elements of machine tool construction and operation relating primarily to their use in manufacturing operations, including:

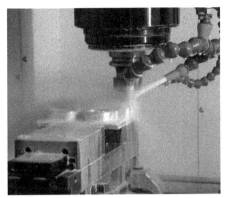

Machining Center

- Work and tool holding elements

- Driving mechanisms that constitute an inherent part of the machine tool

- Components and associated appurtenances

- Nomenclature, designations, sizes, and capacities

- Tests for accuracy of machine tools and of work and tool holding parts or elements

- Movements and adjustments of machine tool elements

- Parts and elements for adjusting, guiding, and aligning work or tools, including slots and tapes, but excluding, punches, dies and screw taps."

ASME B5 Technical Subcommittees

Technical Subcommittees under ASME B5 address the formulation and maintenance of standards in particular disciplines within the scope of the B5 charter. Membership includes a carefully balanced representation in various interest classifications so that no one group dominates. Some examples of the various interest classifications are: users, manufacturers, consultants, insurance interests, universities, testing laboratories, and government regulatory agencies.

There are over 40 Standards that are periodically reviewed to ensure they reflect new developments and technical advances (e.g., new materials, new designs and new applications). Several new standards are under development or being considered in the areas of Tool Holder Interfaces, Machine Performance Evaluation, Micromachining, Energy Assessment, and Robotics. The current list (2015) of B5 subcommittees includes:

TC 11 - Chucks and Chuck Jaws

TC 30 - Power Press Standards

Typical Ball Screw

TC 43 - Ball Screws

TC 45 - Spindle Noses and Tool Shanks for Machining Centers

TC 52 - Machine Tool Performance

TC 54 - Tool Connection Gages for Machine Tools

TC 55 - Tool Identification Systems (RFID)

TC 56 - Information Technology for Machine Tools

TC 65- Micromachining

TC 94 - Cutting Tools

ASME B5 also provides the ANSI-sanctioned Technical Advisory Groups to ISO regarding machine tool technology. These include

Technical Advisory to ISO TC39 Machine Tools

ISO TC39, SC2 – Test conditions for metal cutting machine tools

ISO TC39, SC4 – Woodworking machines

ISO TC39, SC6 – Noise of machine tools

ISO TC39, SC8 – Work holding spindles and chucks

ISO TC39, SC10 – Safety

ISO TC39, WG 7 – Ball screws

ISO TC39, WG9 – Symbols for indications appearing on machine tools

ISO TC39, WG 12 – Environmental evaluation of machine tools

ISO TC39, WG 16 – Production equipment for Microsystems

Technical Advisory to ISO TC29

ISO TC29, SC2 – Cutting tools and their attachments

References

- Altintas, Yusuf. Manufacturing Automation: Metal Cutting Mechanics, Machine Tool Vibrations, and CNC Design. Cambridge University Press, 2000, ISBN 978-0-521-65973-4.

- Cheng, Kai. Machining Dynamics: Fundamentals, Applications and Practices. Springer, 2008, ISBN 978-1-84628-367-3.

- Schmitz, Kai, Tony L., Smith, Scott K. Machining Dynamics: Frequency Response to Improved Productivity. Springer, 2008, ISBN 978-0-387-09644-5.

- Maekawa, Obikawa. Metal Machining: Theory and Applications. Butterworth-Heinemann, 2000, ISBN 978-0-340-69159-5.

- Freitas, Robert A.; Ralph C. Merkle (2004). Kinematic Self-Replicating Machines. Georgetown, Texas: Landes Bioscience. ISBN 1-57059-690-5.

- von Neumann, John (1966). A. Burks, ed. The Theory of Self-reproducing Automata. Urbana, IL: Univ. of Illinois Press. ISBN 0-598-37798-0.

- Raets, Stefan. "Going through the Spin Cycle: Spin by Robert Charles Wilson". tor.com. Retrieved 17 January 2016.

- Anders, Charlie Jane. "The Most Messed Up Book About Robot Consciousness Ever". io9. Retrieved 17 January 2016.

- "Standards & Certification Development Committees". American Society of Mechanical Engineers. 2015. Retrieved 2015-01-28.

6

Numerical Control: An Integrated Study

Numerical control is the automation of any machine tool, it can be operated precisely by commands and the storage medium is usually a computer. Some of the aspects of the numerical control are digital modeling and fabrications, cutter locations, Cartesian co-ordinate robots, CNC routers and CNC plunge millings. This section helps the readers in developing an in depth understanding of the subject matter.

Numerical Control

Computer Numeric Control (CNC) is the automation of machine tools that are operated by precisely programmed commands encoded on a storage medium (computer command module, usually located on the device) as opposed to controlled manually by hand wheels or levers, or mechanically automated by cams alone. Most NC today is computer (or computerized) numerical control (CNC), in which computers play an integral part of the control.

A CNC turning center

In modern CNC systems, end-to-end component design is highly automated using computer-aided design (CAD) and computer-aided manufacturing (CAM) programs. The programs produce a computer file that is interpreted to extract the commands needed to operate a particular machine by use of a post processor, and then loaded into the CNC machines for production. Since any particular component might require the use of a number of different tools – drills, saws, etc. – modern machines often combine multiple tools into a single "cell". In other installations, a number of different machines are used with an external controller and human or robotic operators that move the component from machine to machine. In either case, the series of steps needed to produce any part is highly automated and produces a part that closely matches the original CAD design.

History

The first NC machines were built in the 1940s and 1950s, based on existing tools that were modified with motors that moved the controls to follow points fed into the system on punched tape. These early servomechanisms were rapidly augmented with analog and digital computers, creating the modern CNC machine tools that have revolutionized the machining processes.

Description

Motion is controlled along multiple axes, normally at least two (X and Y), and a tool spindle that moves in the Z (depth). The position of the tool is driven by direct-drive stepper motor or servo motors in order to provide highly accurate movements, or in older designs, motors through a series of step down gears. Open-loop control works as long as the forces are kept small enough and speeds are not too great. On commercial metalworking machines, closed loop controls are standard and required in order to provide the accuracy, speed, and repeatability demanded.

As the controller hardware evolved, the mills themselves also evolved. One change has b10dditional safety interlocks to ensure the operator is far enough from the working piece for safe operation. Most new CNC systems built today are 100% electronically controlled.

CNC-like systems are now used for any process that can be described as a series of movements and operations. These include laser cutting, welding, friction stir welding, ultrasonic welding, flame and plasma cutting, bending, spinning, hole-punching, pinning, gluing, fabric cutting, sewing, tape and fiber placement, routing, picking and placing, and sawing.

Examples of CNC Machines

Mills

CNC mills use computer controls to cut different materials. They are able to translate programs consisting of specific numbers and letters to move the spindle (or workpiece) to various locations and depths. Many use G-code, which is a standardized programming language that many CNC machines understand, while others use proprietary languages created by their manufacturers. These proprietary languages, while often simpler than G-code, are not transferable to other machines. CNC mills have many functions including face milling, shoulder milling, tapping, drilling and some even offer turning. Standard linear CNC mills are limited to 3 axis (X, Y, and Z), but others may also have one or more rotational axes. Today, CNC mills can have 4 to 6 axes.

Lathes

A Tsugami multifunction turn mill machine used for short runs of complex parts.

Lathes are machines that cut workpieces while they are rotated. CNC lathes are able to make fast, precision cuts, generally using indexable tools and drills. They are particularly effective for complicated programs designed to make parts that would be infeasible to make on manual lathes. CNC lathes have similar control specifications to CNC mills and can often read G-code as well as the manufacturer's proprietary programming language. CNC lathes generally have two axes (X and Z), but newer models have more axes, allowing for more advanced jobs to be machined.

Plasma Cutters

CNC plasma cutting

Plasma cutting involves cutting a material using a plasma torch. It is commonly used to cut steel and other metals, but can be used on a variety of materials. In this process, gas (such as compressed air) is blown at high speed out of a nozzle; at the same time an electrical arc is formed through that gas from the nozzle to the surface being cut, turning some of that gas to plasma. The plasma is sufficiently hot to melt the material being cut and moves sufficiently fast to blow molten metal away from the cut.

Electric Discharge Machining

Electric discharge machining (EDM), sometimes colloquially also referred to as spark machining, spark eroding, burning, die sinking, or wire erosion, is a manufacturing process in which a desired shape is obtained using electrical discharges (sparks). Material is removed from the workpiece by a series of rapidly recurring current discharges between two electrodes, separated by a dielectric fluid and subject to an electric voltage. One of the electrodes is called the tool electrode, or simply the "tool" or "electrode," while the other is called the workpiece electrode, or "workpiece."

When the distance between the two electrodes is reduced, the intensity of the electric field in the space between the electrodes becomes greater than the strength of the dielectric (at the nearest point(s)), which electrically break down, allowing current to flow between the two electrodes. This phenomenon is the same as the breakdown of a capacitor. As a result, material is removed from both the electrodes. Once the current flow stops (or it is stopped – depending on the type of generator), new liquid dielectric is usually conveyed into the inter-electrode volume, enabling the solid particles (debris) to be carried away and the insulating properties of the dielectric to be restored. Adding new liquid dielectric in the inter-electrode volume is commonly referred to as flushing. Also, after a current flow, a difference of potential between the two electrodes is restored to what it was before the breakdown, so that a new liquid dielectric breakdown can occur.

Wire EDM

Also known as wire cutting EDM, wire burning EDM, or traveling wire EDM, this process uses spark erosion to machine or remove material with a traveling wire electrode from any electrically conductive material. The wire electrode usually consists of brass or zinc-coated brass material.

Sinker EDM

Sinker EDM, also called cavity type EDM or volume EDM, consists of an electrode and workpiece submerged in an insulating liquid—often oil but sometimes other dielectric fluids. The electrode and workpiece are connected to a suitable power supply, which generates an electrical potential between the two parts. As the electrode approaches the workpiece, dielectric breakdown occurs in the fluid forming a plasma channel) and a small spark jumps.

Water Jet Cutters

A water jet cutter, also known as a waterjet, is a tool capable of slicing into metal or other materials (such as granite) by using a jet of water at high velocity and pressure, or a mixture of water and an abrasive substance, such as sand. It is often used during fabrication or manufacture of parts for machinery and other devices. Waterjet is the preferred method when the materials being cut are sensitive to the high temperatures generated by other methods. It has found applications in a diverse number of industries from mining to aerospace where it is used for operations such as cutting, shaping, carving, and reaming.

Other CNC Tools

Many other tools have CNC variants, including:

- Drills
- EDMs
- Embroidery machines
- Lathes
- Milling machines
- Canned cycle
- Wood routers
- Sheet metal works (Turret punch)
- Wire bending machines
- Hot-wire foam cutters
- Plasma cutters
- Water jet cutters
- Laser cutting
- Oxy-fuel
- Surface grinders
- Cylindrical grinders
- 3D Printing
- Induction hardening machines
- Submerged welding
- Knife cutting
- Glass cutting

Tool / Machine Crashing

In CNC, a "crash" occurs when the machine moves in such a way that is harmful to the machine, tools, or parts being machined, sometimes resulting in bending or breakage of cutting tools, accessory clamps, vises, and fixtures, or causing damage to the machine itself by bending guide rails, breaking drive screws, or causing structural components to crack or deform under strain. A mild crash may not damage the machine or tools, but may damage the part being machined so that it must be scrapped.

Many CNC tools have no inherent sense of the absolute position of the table or tools when turned on. They must be manually "homed" or "zeroed" to have any reference to work from, and these limits are just for figuring out the location of the part to work with it, and aren't really any sort of hard motion limit on the mechanism. It is often possible to drive the machine outside the physical bounds of its drive mechanism, resulting in a collision with itself or damage to the drive mechanism. Many machines implement control parameters limiting axis motion past a certain limit in addition to physical limit switches. However, these parameters can often be changed by the operator.

Many CNC tools also don't know anything about their working environment. Machines may have load sensing systems on spindle and axis drives, but some do not. They blind-ly follow the machining code provided and it is up to an operator to detect if a crash is either occurring or about to occur, and for the operator to manually abort the cutting process. Machines equipped with load sensors can stop axis or spindle movement in response to an overload condition, but this does not prevent a crash from occurring. It may only limit the damage resulting from the crash. Some crashes may not ever over-load any axis or spindle drives.

If the drive system is weaker than the machine structural integrity, then the drive sys-tem simply pushes against the obstruction and the drive motors "slip in place". The machine tool may not detect the collision or the slipping, so for example the tool should now be at 210 mm on the X axis, but is, in fact, at 32mm where it hit the obstruction and kept slipping. All of the next tool motions will be off by −178mm on the X axis, and all future motions are now invalid, which may result in further collisions with clamps, vis-es, or the machine itself. This is common in open loop stepper systems, but is not pos-sible in closed loop systems unless mechanical slippage between the motor and drive mechanism has occurred. Instead, in a closed loop system, the machine will continue to attempt to move against the load until either the drive motor goes into an overcurrent condition or a servo following error alarm is generated.

Collision detection and avoidance is possible, through the use of absolute position sen-sors (optical encoder strips or disks) to verify that motion occurred, or torque sensors or power-draw sensors on the drive system to detect abnormal strain when the ma-chine should just be moving and not cutting, but these are not a common component of most hobby CNC tools.

Instead, most hobby CNC tools simply rely on the assumed accuracy of stepper motors that rotate a specific number of degrees in response to magnetic field changes. It is often assumed the stepper is perfectly accurate and never missteps, so tool position monitoring simply involves counting the number of pulses sent to the stepper over time. An alternate means of stepper position monitoring is usually not available, so crash or slip detection is not possible.

Commercial CNC metalworking machines use closed loop feedback controls for axis movement. In a closed loop system, the control is aware of the actual position of the axis at all times. With proper control programming, this will reduce the possibility of a crash, but it is still up to the operator and programmer to ensure that the machine is operated in a safe manner. However, during the 2000s and 2010s, the software for machining simulation has been maturing rapidly, and it is no longer uncommon for the entire machine tool envelope (including all axes, spindles, chucks, turrets, tool holders, tailstocks, fixtures, clamps, and stock) to be modeled accurately with 3D solid models, which allows the simulation software to predict fairly accurately whether a cycle will involve a crash. Although such simulation is not new, its accuracy and market penetration are changing considerably because of computing advancements.

Numerical Precision and Equipment Backlash

Within the numerical systems of CNC programming it is possible for the code generator to assume that the controlled mechanism is always perfectly accurate, or that precision tolerances are identical for all cutting or movement directions. This is not always a true condition of CNC tools. CNC tools with a large amount of mechanical backlash can still be highly precise if the drive or cutting mechanism is only driven so as to apply cutting force from one direction, and all driving systems are pressed tightly together in that one cutting direction. However a CNC device with high backlash and a dull cutting tool can lead to cutter chatter and possible workpiece gouging. Backlash also affects precision of some operations involving axis movement reversals during cutting, such as the milling of a circle, where axis motion is sinusoidal. However, this can be compensated for if the amount of backlash is precisely known by linear encoders or manual measurement.

The high backlash mechanism itself is not necessarily relied on to be repeatedly precise for the cutting process, but some other reference object or precision surface may be used to zero the mechanism, by tightly applying pressure against the reference and setting that as the zero reference for all following CNC-encoded motions. This is similar to the manual machine tool method of clamping a micrometer onto a reference beam and adjusting the Vernier dial to zero using that object as the reference.

Positioning Control System in NC

In NC system the position of the tool is defined by the part program of instruction that store by the machine control unit(MCU). Two type of positioning control system are used in the NC system:

1. Open loop control system

2. closed loop control system

Open Loop Control System

This system operates without verifying that actual position achieved by the worktable. After the executing the program by MCU, it does not use any feedback so it known as the open loop system.

Closed Loop Control System

A closed loop control system uses feedback measurements to confirm that the final position of the worktable is the location specified in the program.

Comparison of Some Parameter of The Open Loop and Closed Loop System

S. No.	Open loop control system	closed loop control system
1.	Less expensive	More expensive
2.	Not use any feedback mechanism	Use feedback mechanism
3.	Any type of motion control system	Mostly used for continuous path system
4.	Generally used stepper motor	Servo motor
5.	Less accurate at high speed	Accuracy is high compared to open loop system

Digital Modeling and Fabrication

Digital modeling and fabrication is a process that joins design with production through the use of 3D modeling software or computer-aided design (CAD) and additive and subtractive manufacturing processes. 3D printing falls under additive, while machining falls under subtractive. These tools allow designers to produce material digitally, which is something greater than an image on screen, and actually tests the accuracy of the software and computer lines.

Computer milling and fabrication integrate the computer assisted designs with that of the construction industry. In this process, the sequence of operations becomes the critical characteristic in procedure. Architects can propose complex surfaces, where the properties of materials should push the design.

Modeling

Digitally fabricated objects are created with a variety of CAD software packages, using both 2D vector drawing, and 3D modeling. Types of 3D models include wireframe, solid, surface and mesh. A design is one or more of these model types.

Machines for Fabrication

CNC Router

CNC stands computer numerical control. CNC mills or routers include proprietary software which interprets 2D vector drawings or 3D models and converts this information to a G-code, which represents specific CNC functions in alphanumeric format which the CNC mill can interpret. The g-codes drive a machine tool, a powered mechanical device typically used to fabricate components. CNC machines are classified according to the number of axes that they possess, with 3, 4, and 5 axis machines all being common, and industrial robots being described with having as many as 9 axes. CNC machines are specifically successful in milling materials such as plywood, foam board, and metal at a fast speed. CNC machine beds are typically large enough to allow 4' × 8' (123 cm x 246 cm) sheets of material, including foam several inches thick, to be cut.

Laser Cutter

The laser cutter is a machine that uses a laser to cut materials such as chip board, matte board, felt, wood, and acrylic up to 3/8" (1 cm) thickness. The laser cutter is often bundled with a driver software which interprets vector drawings produced by any number of CAD software platforms.

The laser cutter is able to modulate the speed of the laser head, as well as the intensity and resolution of the laser beam, and as such is able both cut and score material, as well as approximate raster graphics.

Objects cut out of materials can be used in the fabrication of physical models, which will only require the assembly of the flat parts.

3D Printers

3d printers use a variety of methods and technology to assemble physical versions of digital objects. Typically desktop 3d printers can make small plastic 3d objects. They use a roll of thin plastic filament, melting the plastic and then depositing it precisely to cool and harden. They normally build 3D objects from bottom to top in a series of many very thin plastic horizontal layers. This process often happens over the course of a several hours.

Fused deposition modeling, also known as fused filament fabrication, uses a 3-axis robotic system that extrudes material, typically a thermoplastic, one thin layer at a time and progressively builds up a shape. Examples of machines that use this method are the Dimension 768 and the Ultimaker.

Stereolithography uses a high intensity light projector, usually using DLP technology, with a photosensitive polymer resin. It will project the profile of an object to build a single layer, curing the resin into a solid shape. Then the printer will move the object out of the way by a small amount and project the profile of the next layer. Examples of devices that use this method are the Form-One printer and Os-RC Illios.

Selective laser sintering uses a laser to trace out the shape of an object in a bed of finely powdered material that can be fused together by application of heat from the laser. After one layer has been traced by a laser, the bed and partially finished part is moved out of the way, a thin layer of the powdered material is spread, and the process is repeated. Typical materials used are alumide, steel, glass, thermoplastics (especially nylon), and certain ceramics. Example devices include the Formiga P 110 and the Eos EosINT P730.

Powder printers work in a similar manner to SLS machines, and typically use powders that can be cured, hardened, or otherwise made solid by the application of a liquid binder that is delivered via an inkjet printhead. Common materials are plaster of paris, clay, powdered sugar, wood-filler bonding putty, and flour, which are typically cured with water, alcohol, vinegar, or some combination thereof. The major advantage of powder and SLS machines is their ability to continuously support all parts of their objects throughout the printing process with unprinted powder. This permits the assembly of geometries not easily otherwise created. However, these printers are often more complex and expensive. Examples of printers using this method are the ZCorp Zprint 400 and 450.

Cutter Location

A cutter location (CLData) refers to the position which a CNC milling machine has been instructed to hold a milling cutter by the instructions in the program (typically G-code).

Each line of motion controlling G-code consists of two parts: the type of motion from the last cutter location to the next cutter location (e.g. "G01" means linear, "G02" means circular), and the next cutter location itself (the cartesian point (20, 1.3, 4.409) in this example). "G01 X20Y1.3Z4.409"

The fundamental basis for creating the cutter paths suitable for CNC milling are functions that can find valid cutter locations, and stringing them together in a series.

There are two broad and conflicting approaches to the problem of generating valid cutter locations, given a CAD model and a tool definition: calculation by offsets, and calculation against triangles.

The most common example of the general cutter location problem is cutter radius compensation (CRC), in which an endmill (whether square end, ball end, or bull end) must be offset to compensate for its radius.

Since the 1950s, CRC calculations finding tangency points on the fly have been done automatically within CNC controls, following the instructions of G-codes such as G40, G41, and G42. The chief inputs have been the radius offset values stored in the offset registers (typically called via address D) and the left/right climb/conventional distinction called via G41 or G42 (respectively). With the advent of CAM software, which added a software-aided option to complement the older manual-programming environment, much of the CRC calculations could be moved to the CAM side, and various modes could be offered for how to handle CRC.

Although 2-axis or 2.5-axis CRC problems (such as calculating toolpaths for a simple profile in the XY plane) are quite simple in terms of computational power, it is in the 3-, 4-, and 5-axis situations of contouring 3D objects with a ball-endmill that CRC becomes rather complex. This is where CAM becomes especially vital and far outshines manual programming. Typically the CAM vector output is postprocessed into G-code by a postprocessor program that is tailored to the particular CNC control model. Some late-model CNC controls accept the vector output directly, and do the translation to servo inputs themselves, internally.

Cutter Location by Offsets

Start with a UV parametric point in a freeform surface, calculate the xyz point and the normal, and offset from the point along the normal in a way consistent with the tool definition so that the cutter is now tangent to the surface at that point.

Problems: may collide or gouge with the model elsewhere, and there is no way to tell this is happening except with a full implementation of the triangulated approach.

Most published academics believe this is the way to find cutter locations, and that the problem of collisions away from the point of contact is soluble. However, nothing printed so far comes close to handling real world cases.

Cutter Location Against Triangles

Start with the XY component for a cutter location and loop across every triangle in the model. For each triangle which crosses under the circular shadow of the cutter, calculate the Z value of the cutter location required for it to exactly touch the triangle, and find the maximum of all such values. Hwang et al. describe this approach in 1998, for cylindrical, ball-end, and bull-end milling tools. These ideas are further developed in a 2002 paper by Chuang et al. In a paper from 2004 Yau et al. describe an algorithm for locating an APT-cutter against triangles. Yau et al. use a kd-tree for finding overlapping triangles.

Problems: requires a lot of memory to hold enough triangles to register the model at a tight enough tolerance, and it takes longer to program to get your initial cutter location values. However, they are at least guaranteed valid in all cases.

This is how all major CAM systems do it these days because it works without failing no matter what the complexity and geometry of the model, and can be made fast later. Reliability is far more important than efficiency.

The above refers to 3-axis machines. 5-axis machines need a special entry of their own.

ZMap

The ZMap algorithm was proposed in the academic literature by Byoung K Choi in 2003 as a way of precalculating and storing a regular array of Cutter Location values in the computer memory. The result is a model of the height map of cutter positions from which in between values can be interpolated.

Due to accuracy issues, this was generalized into an Extended ZMap, or EZMap, by the placement of "floating" points in between the fixed ZMap points. The location of the EZMap points are found iteratively when the ZMap is created. EZMap points are only placed where sharp edges occur between the normal ZMap points; a completely flat source geometry will not require any EZMap points.

Cartesian Coordinate Robot

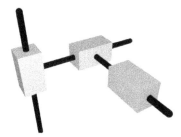

Kinematic diagram of cartesian (coordinate) robot

a plotter is an implemention of the cartesian coordinate robot

A cartesian coordinate robot (also called linear robot) is an industrial robot whose three principal axis of control are linear (i.e. they move in a straight line rather than rotate) and are at right angles to each other.The three sliding joints correspond to moving the wrist up-down,in-out,back-forth. Among other advantages, this mechanical arrangement simplifies the Robot control arm solution. Cartesian coordinate robots with the horizontal member supported at both ends are sometimes called Gantry robots; mechanically, they resemble gantry cranes, although the latter are not generally robots. Gantry robots are often quite large.

A popular application for this type of robot is a computer numerical control machine (CNC machine) and 3D printing. The simplest application is used in milling and drawing machines where a pen or router translates across an x-y plane while a tool is raised and lowered onto a surface to create a precise design. Pick and place machines and plotters are also based on the principal of the cartesian coordinate robot.

CNC Router

A CNC router (Or Computer Numerical Control router) is a computer-controlled cutting machine related to the hand held router used for cutting various hard materials, such as wood, composites, aluminium, steel, plastics, and foams. CNC stands for *computer numerical control*. CNC routers can perform the tasks of many carpentry shop machines such as the panel saw, the spindle moulder, and the boring machine. They can also cut mortises and tenons.

A CNC router is very similar in concept to a CNC milling machine. Instead of routing by hand, tool paths are controlled via computer numerical control. The CNC router is one of many kinds of tools that have CNC variants.

A CNC router typically produces consistent and high-quality work and improves factory productivity. Unlike a jig router, the CNC router can produce a one-off as effectively as repeated identical production. Automation and precision are the key benefits of cnc router tables.

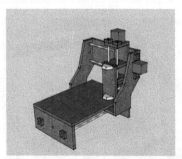

Drawing of a Tabletop DIY - CNC router. Silver: Iron, Red: Stepper Motors, Light Brown: MDF, Dark Brown: Hard Wood

A CNC router can reduce waste, frequency of errors, and the time the finished product takes to get to market.

Applications

A CNC router can be used in the production of many different items, such as door carvings, interior and exterior decorations, wood panels, sign boards, wooden frames, moldings, musical instruments, furniture, and so on. In addition, the CNC router helps in the thermoforming of plastics by automating the trimming process. CNC routers can help ensure part repeatability and sufficient factory output.

Overview of CAM (Computer-aided Manufacturing)

CAM software makes the CAD drawing/design into a code called g-code.This code the CNC machine can understand. In short, CNC technology is not very complicated. It is a tool controlled by a computer. It only becomes more sophisticated when considering how the computer controls the tool. The illustration shows what a bare bones CNC machine might look like without its controller.

Sizes and Configurations of Router

CNC routers come in many configurations, from small home-style D.I.Y. "desktop" like k2 cnc, to large industrialCNC routers used in sign shops, cabinet making, aerospace and boat-making facilities. Originally CNC routers added computer control to consumer router power tools. Although there are many configurations, most CNC routrs have a few specific parts: a dedicated CNC controller, one or more spindle motors, servo motors, Stepper Motors, servo amplifiers, AC inverter frequency drives, linear guides, ball screws and a workspace bed or table. In addition, CNC routers may have vacuum pumps, with grid table tops or t slot hold down fixtures to hold the parts in place for cutting. CNC routers are generally available in 3-axis and 5-axis CNC formats. Many Manufacturers offer A and B Axis for full 5 Axis capabilities and rotary 4th axis.

Controlling The Machine

The CNC router is controlled by a computer. Coordinates are uploaded into the machine controller from a separate CAD program. CNC router owners often have two software applications—one program to make designs (CAD) and another to translate those designs into a 'G-Code' program of instructions for the machine (CAM). As with CNC milling machines, CNC routers can be controlled directly by manual programming, and CAD/CAM opens up wider possibilities for contouring, speeding up the programming process and in some cases creating programs whose manual programming would be, if not truly impossible, certainly commercially impractical.

Types

Wood

A CNC wood router is a CNC Router tool that creates objects from wood. CNC stands for computer numerical control. The CNC works on the Cartesian coordinate system (X, Y, Z) for 3D motion control. Parts of a project can be designed in the computer with a CAD/CAM program, and then cut automatically using a router or other cutters to produce a finished part.The CNC Router is ideal for hobbies, engineering prototyping, product development, art, and production work.

A typical CNC wood router

Metal

Milling is the machining process of using rotary cutters to remove material from a workpiece advancing (or *feeding*) in a direction at an angle with the axis of the tool. It covers a wide variety of different operations and machines, on scales from small individual parts to large, heavy-duty gang milling operations. It is one of the most commonly used processes in industry and machine shops today for machining parts to precise sizes and shapes.

Stone

A stone CNC router is a type of CNC router machine for marble, granite, artificial stone, tombstone, ceramic tiles, glass carving, engraving, cutting, polishing as arts and crafts, pictures, etc.

CNC Riveting

CNC riveting is a CNC process used for obtaining permanent mechanical fastening of geometrical shapes ranging from a simple to complex shape like fuselage of an aircraft. This is done in a shorter duration of time with a high riveting rate. The process is fast robust and is flexible in nature thus improving its usage and providing reliability to the riveted joint along with the final product quality.CNC riveting can be used for a variety of operations ranging from riveting and fastening the belts, skin panels, shear ties, and other internal fuselage component.

The CNC Riveting machines generally consists of a solid frame made by welding steel and protective aluminium frames used for protection fitted with polycarbonate panes.The dynamic drive of the coordinates axes is achieved via recirculating ball and screw,servo motors and motion control units makes the high speed movement possible.For the mounting of the riveting units solid C Frames are used. The riveting program can be given various parameters and these can be changed or altered as desired being cnc programs as per the requirement.

CNC Riveting Machine Variants

CNC riveting machines has many variants.

CNC Duct Riveting Cell

A CNC controlled automatic riveting work cell with a knee type design drill machine has sixty inch throat depth with four position upper head. this machine can apply sixteen pounds of upset force. This machine is equipped with a dual drill spindles,one for carrying out drilling and other for deburring. it was essentially developed for fabrication of tubular assemblies which are fed into the throat of the machine over an eight-inch square lower knee.The CNC controlled four axis positioning system presents the part to the machine for carrying out the riveting.

CNC Riveting Machines With Stationary Machine Table

These CNC machines act as stand alone workstations for heavy and bigger workpieces. they are simple in design and require work holding fixtures only. thus the clamping and component query devices do not cost much.

CNC Riveting Machines With Indexing Tables

These special purpose CNC riveting machines with different sizes of indexing tables can be composed of different coordinate axes and riveting machines.These machines have different versions for the particular application and are configured accordingly. their frame is made of steel and protective shells and frames are made of aluminium fitted with polycarbonate panes aiding in visibility of riveting operation.the coordinate system is with linear units and recirculating ball screws, index table is electrically operated rotary indexing table with brake motor.It has two or four fixed indexing stations, the index table are NC flexible rotary indexing tables which are actuated by two hand control or by a pedal switch. The machine has an automatic tool changer.

CNC Riveting Machines With Transfer System

These CNC riveting machines are made for the use in manufacturing lines.The fixtures are coded and interfaces are customized to make the connection of several CNC riveting machines and linking them with other manufacturing systems making it possible to obtaining a high degree of automation.

Multiple Axes CNC Riveting Cell

This is the latest technological development in automated fastening.This technology is versatile and can be used for riveting of high curvature fuselage panels to low curvature wing panels,bulkheads,floor etc. The tooling changeover is minimized thus part throughput is maximized.

Advantages

The main advantages of this type of CNC riveting machine is that it can use a minimum distance between rivets and rivets of different length or heights can be used .High flexibility and changeover time due to programmable memory.It can process many workpieces and different rivets can be used in one operation.Picking and placing operations are done in parallel with the primary operation time thus saving cost.Ergonomics of work station.Menu based navigation makes the programming lucid. High acceleration rate with high positioning accuracy. The movement speed is up to 400 mm/s

CNC Wood Router

A CNC wood router is a CNC router tool that creates objects from wood. CNC stands for *computer numerical control*. The CNC works on the Cartesian coordinate system (X, Y, Z) for 3D motion control. Parts of a project can be designed in the computer with a CAD/CAM program, and then cut automatically using a router or other cutters to produce a finished part.

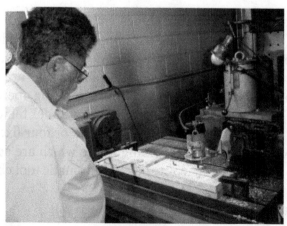

A CNC metalworking machine with a wood router attached to it, turning it into a makeshift CNC router. Cutting bit rotation speeds on metal working equipment is typically too slow to produce good results in wood.

Typical work done by a CNC wood router

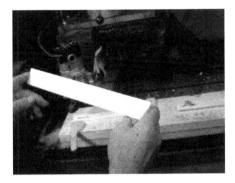
Typical wood piece before router cutting

The CNC router is ideal for hobbies, engineering prototyping, product development, art, and production work.

Operation

A CNC wood router uses CNC (computer numerical control) and is similar to a metal CNC mill with the following differences:

- The wood router typically spins faster — with a range of 13,000 to 24,000 RPM

- Professional quality machines frequently use surface facing tools up to 3" in diameter or more, and spindle power from 5 to 15 horsepower. Machines capable of routing heavy material at over a thousand inches per minute are common.

- Some machines use smaller toolholders MK2 (Morse taper #2 - on older machines), ISO-30, HSK-63 or the tools just get held in a collet tool holder affixed directly to the spindle nose. ISO-30 and HSK-63 are rapid-change toolholding systems. HSK-63 has begun to supplant the ISO-30 as the rapid change standard in recent years.

A wood router is controlled in the same way as a metal mill, but there are CAM and CAD applications such as Artcam, Mastercam, Bobcad, and AlphaCam, which are specifically designed for use with wood routers.

Wood routers are frequently used to machine other soft materials such as plastics.

Typical three-axis CNC wood routers are generally much bigger than their metal shop counterparts. 5' x 5', 4' x 8', and 5' x 10' are typical bed sizes for wood routers. They can be built to accommodate very large sizes up to, but not limited to 12' x 100'. The table can move, allowing for true three axis (xyz) motion, or the gantry can move, which requires the third axis to be controlled by two slaved servo motors.

Features

Separate Heads

Some wood routers have multiple separate heads that can come down simultaneously or not. Some routers have multiple heads that can run complete separate programs on separate tables all while being controlled by the same interface.

Dust Collection / Vacuum Collector

The wood router typically has 6"-10" air ducts to suck up the wood chips/dust created. They can be piped to a stand-alone or full shop dust collection system.

Some wood routers are specialized for cabinetry and have many drills that can be programmed to come down separately or together. The drills are generally spaced 32 mm apart on centres - a spacing system called 32 mm System. This is for the proper spacing of shelving for cabinets. Drilling can be vertical or horizontal (in the Y or X axis from either side/end of the workpiece) which allows a panel to be drilled on all four edges as well as the top surface. Many of these machines with large drilling arrays are derived from CNC point-to-point borers.

Securing The Workpiece

Suction Systems

The wood router typically holds wood with suction through the table or pods that raise the work above the table. Pods may be used for components which require edge profiling (or undercutting), are manufactured from solid wood or where greater flexibility in production is required. This type of bed requires less extraction with greater absolute vacuum.

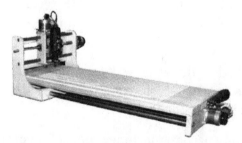

A typical CNC wood router with suction holes visible

A second type hold down uses a spoil board. This allows vacuum suction through a low density table and allows the placement of parts anywhere on the table. These types of tables are typically used for nest-based manufacturing (NBM) where multiple components are routed from a single sheet. This type of manufacturing precludes edge drilling or undercut edge work on components.

Vacuum pumps are required with both types of tables where volume and "strength" are determined based on the types of materials being cut.

CNC Plunge Milling

CNC plunge milling is sometimes called z-axis milling also. In this process, the feed is provided linearly along the tool axis while doing CNC processing.

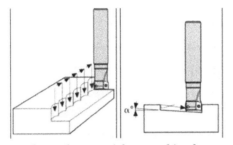

This image shows the material removal in plunge milling.

Plunge milling is very good method for rough machining process of complex shape or free form shapes like impeller parts. In multi axis plunge milling, the optimization of plunge cutter section selection and generating the tool path for free form surface is very important to improve the efficiency and effectiveness. In plunge milling, after each plunge the milling cutter is offset by some value and then the material surface is removed in the form of lunula. The material removal rate is computed by area of lunula and the feed rate. At the entry and exit of milling cutter, the radial offset has not any influence on the condition of surface. At the maximum cutting velocity, the surface obtained is clean whatever the feed rate per tooth on entry but on exit the high value of feed rate gives the deteriorated surface. The surface roughness value always increases with feed rate in plunge milling. The simulation of dynamic uncut chip thickness which is generated by plunge milling can be done by tracking the position of plunge cutter center. This simulation shows the regenerative effect with variation of phase difference. Then the model of uncut chip thickness and cutting force coefficient with cutting edge radius are entered into time domain model. Finally, with the help of time domain solution the stability of machine and vibrations are estimated. The cutting parameters play a key role in plunge milling. The cutting force and machine stability both are influenced by machining parameters. Frequency domain model can be used to estimate the machining stability.

Advantages of CNC Plunge Milling

The plunge milling has following advantage over conventional milling-

1. The radial cutting forces, which is responsible for the deformation of the tool

and the work-piece, is very small.

2. The material which is difficult to cut can be rough machined easily.

3. Comparing to conventional milling,in CNC plunge milling the feed per tooth is less.

4. The vibration can be avoided in the machine so that is why it is suitable for deep cavity machining like mould and cavity making. In deep milling the overhang of large length tool is also adjusted.

References

- P. GROOVER, MIKELL (2015). AUTOMATION, PRODUCTION SYSTEM, AND COMPUTER-INTEGRAE MANUFACTURING. india: PEARSON. pp. 166–167. ISBN 9789332549814.

- Lynch, Mike. "Five CNC Myths and Misconceptions | Modern Machine Shop". www.mmsonline.com. Retrieved 2016-02-17.

- Industries, Precision Metal. "Automated Laser Cutting | Precision Metal Industries". www.pmi-quality.com. Retrieved 2016-02-17.

- E. Endres, Thomas E. Endres. "CNC Duct Riveting Work Cell". www.sae.org. Thomas E. Endres. Retrieved 16 March 2015.

- "Computerized Numerical Control". www.sheltonstate.edu. Shelton State Community College. Retrieved March 24, 2015.

- Mike Lynch, "Key CNC Concept #1—The Fundamentals Of CNC", Modern Machine Shop, 4 January 1997. Accessed 11 February 2015.

7

Industrial Revolution

One of the most important technological developments of recent human history is the Industrial Revolution. It greatly developed the mechanization process, and one of the important inventions during the Industrial Revolution is the steam engine. Steam engines were not only used in the mining industry but also in various other industrial settings. Interchangeable parts, paper machine and coal mining are the topics discussed in the following chapter.

Steam Power During the Industrial Revolution

The steam engine was one of the most important technologies of the Industrial Revolution, although steam did not replace water power in importance in Britain until after the Industrial Revolution. From Englishman Thomas Newcomen's atmospheric engine, of 1712, through major developments by Scottish inventor and mechanical engineer James Watt, the steam engine began to be used in many industrial settings, not just in mining, where the first engines had been used to pump water from deep workings. Early mills had run successfully with water power, but by using a steam engine a factory could be located anywhere, not just close to water. Water power varied with the seasons and was not available at times due to freezing, floods and dry spells.

In 1775 Watt formed an engine-building and engineering partnership with manufacturer Matthew Boulton. The partnership of Boulton & Watt became one of the most important businesses of the Industrial Revolution and served as a kind of creative technical centre for much of the British economy. The partners solved technical problems and spread the solutions to other companies. Similar firms did the same thing in other industries and were especially important in the machine tool industry. These interactions between companies were important because they reduced the amount of research time and expense that each business had to spend working with its own resources. The technological advances of the Industrial Revolution happened more quickly because firms often shared information, which they then could use to create new techniques or products.

From mines to mills, steam engines found many uses in a variety of industries. The introduction of steam engines improved productivity and technology, and allowed the creation of smaller and better engines. After Richard Trevithick's development of the high-pressure engine, transport-applications became possible, and steam engines found their way into boats, railways, farms and road vehicles. Steam engines are an example of how changes brought by industrialization led to even more changes in other areas.

The development of the stationary steam engine was an essential early element of the Industrial Revolution, however it should be remembered that for most of the period of the Industrial Revolution the majority of industries still relied on wind and water power as well as horse and man-power for driving small machines.

Thomas Savery's Steam Pump

The industrial use of steam power started with Thomas Savery in 1698. He constructed and patented in London the first engine, which he called the "Miner's Friend" since he intended it to pump water from mines. This machine used steam at 8 to 10 atmospheres (120–150 psi) and had no moving parts other than hand-operated valves. The steam once admitted into the cylinder was first condensed by an external cold water spray, thus creating a partial vacuum which drew water up through a pipe from a lower level; then valves were opened and closed and a fresh charge of steam applied directly on to the surface of the water now in the cylinder, forcing it up an outlet pipe discharging at higher level. The engine was used as a low-lift water pump in a few mines and numerous water works, but it was not a success since it was limited in pumping height and prone to boiler explosions.

Thomas Newcomen's Steam Engine

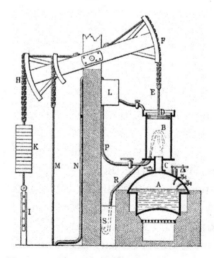

Newcomen's atmospheric steam engine

The first safe and successful steam power plant was introduced by Thomas Newcomen in 1712. Newcomen apparently conceived his machine quite independently of Savery, but as the latter had taken out a very wide-ranging patent, Newcomen and his associates were obliged to come to an arrangement with him, marketing the engine until 1733 under a joint patent. Newcomen's engine appears to have been based on Papin's experiments carried out 30 years earlier, and employed a piston and cylinder, one end of which was open to the atmosphere above the piston. Steam just above atmospheric pressure (all that the boiler could stand) was introduced into the lower half of the cylinder beneath the piston during the gravity-induced upstroke; the steam was then condensed by a jet of cold water injected into the steam space to produce a partial vacuum; the pressure differential between the atmosphere and the vacuum on either side of the piston displaced it downwards into the cylinder, raising the opposite end of a rocking beam to which was attached a gang of gravity-actuated reciprocating force pumps housed in the mineshaft. The engine's downward power stroke raised the pump, priming it and preparing the pumping stroke. At first the phases were controlled by hand, but within ten years an escapement mechanism had been devised worked by of a vertical *plug tree* suspended from the rocking beam which rendered the engine self-acting.

A number of Newcomen engines were successfully put to use in Britain for draining hitherto unworkable deep mines, with the engine on the surface; these were large machines, requiring a lot of capital to build, and produced about 5 hp. They were extremely inefficient by modern standards, but when located where coal was cheap at pit heads, opened up a great expansion in coal mining by allowing mines to go deeper. Despite their disadvantages, Newcomen engines were reliable and easy to maintain and continued to be used in the coalfields until the early decades of the nineteenth century. By 1729, when Newcomen died, his engines had spread to France, Germany, Austria, Hungary and Sweden. A total of 110 are known to have been built by 1733 when the joint patent expired, of which 14 were abroad. In the 1770s, the engineer John Smeaton built some very large examples and introduced a number of improvements. A total of 1,454 engines had been built by 1800.

James Watt's Steam Engines

A fundamental change in working principles was brought about by James Watt. With the close collaboration of Matthew Boulton, he had succeeded by 1778 in perfecting his steam engine which incorporated a series of radical improvements, notably, the use of a steam jacket around the cylinder to keep it at the temperature of the steam and, most importantly, a steam condenser chamber separate from the piston chamber. These improvements increased engine efficiency by a factor of about five, saving 75% on coal costs.

The Newcomen engine could not, at the time, be easily adapted to drive a rotating wheel, although Wasborough and Pickard did succeed in doing so in about 1780. However, by 1783 the more economical Watt steam engine had been fully developed into a double-acting rotative type with a centrifugal governor, parallel motion and flywheel which meant that it could be used to directly drive the rotary machinery of a factory or mill. Both of Watt's basic engine types were commercially very successful.

By 1800, the firm Boulton & Watt had constructed 496 engines, with 164 driving reciprocating pumps, 24 serving blast furnaces, and 308 powering mill machinery; most of the engines generated from 5 to 10 hp. An estimate of the total power that could be produced by all these engines was about 11,200 hp. This was still only a small fraction of the total power generated in Britain by waterwheels (120,000 hp) and by windmills (15,000 hp). Newcomen and other steam engines generated at the same time about 24,000 hp.

Development After Watt

The development of machine tools, such as the lathe, planing and shaping machines powered by these engines, enabled all the metal parts of the engines to be easily and accurately cut and in turn made it possible to build larger and more powerful engines.

In the early 19th century after the expiration of Watt's patent, the steam engine underwent great increases in power due to the use of higher pressure steam which Watt had always avoided because of the danger of exploding boilers, which were in a very primitive state of development.

Until about 1800, the most common pattern of steam engine was the beam engine, built as an integral part of a stone or brick engine-house, but soon various patterns of self-contained portative engines (readily removable, but not on wheels) were developed, such as the table engine. Further decrease in size due to use of higher pressure came towards the end of the 18th Century when the Cornish engineer, Richard Trevithick and the American engineer, Oliver Evans, independently began to construct higher pressure (about 40 pounds per square inch (2.7 atm)) engines which exhausted into the atmosphere. This allowed an engine and boiler to be combined into a single unit compact and light enough to be used on mobile road and rail locomotives and steam boats.

Trevithick was a man of versatile talents, and his activities were not confined to small applications. Trevithick developed his large Cornish boiler with an internal flue from about 1812. These were also employed when upgrading a number of Watt pumping engines, greatly increasing power and productivity; this led to the highly efficient large Cornish engines that continued to be built right up to the end of the 19th Century.

The Corliss Engine

The Corliss Engine displayed at the International Exhibition of Arts, Manufactures and Products of the Soil and Mine of 1876

Due to the ever-increasing power demands of the 1800s, especially in manufacturing, innovations were made to existing steam engines and a number of entirely new steam engines were developed. Of these, few brought the high levels of horsepower and efficiency produced by the Corliss engine. Named after its inventor, George Henry Corliss, this stationary steam engine was introduced to the world in 1849. The engine boasted a number of desired features, including fuel efficiency (lowering cost of fuel by a third or more), low maintenance costs, 30% higher rate of power production, high thermal efficiency, and the ability to operate under light, heavy, or varying loads while maintaining high velocity and constant speed. While the engine was loosely based on existing steam engines keeping the simple piston-flywheel design, the majority of these features were brought about by the engine's unique valves and valve gears. Unlike most engines employed during the era that were using mainly slide-valve gears, Corliss created his own system that used a wrist plate to control a number of different valves. Each cylinder was equipped with four valves, with exhaust and inlet valves at both ends of the cylinder. Through a precisely tuned series of events opening and closing these valves, steam is admitted and released at a precise rate allowing for linear piston motion. This provided the engine's most notable feature, the automatic variable cut-off mechanism. This mechanism is what allowed the engine to function under varying loads without losing efficiency, stalling, or being damaged. Using a series of cam gears, which could adjust valve timing (essentially acting as a throttle), the engine's speed and horsepower was adjusted. This proved extremely useful for most of the engine's applications. In the textile industry, it allowed for production at much higher speeds while lowering the likelihood that threads would break. In metallurgy, the extreme and abrupt variations of load experienced in rolling mills were also countered by the technology. These examples demonstrate that the Corliss engine was able to lead to much higher rates of production, while preventing costly damages to machinery and materials. It was referred to as "the most perfect regulation of speed."

Corliss kept a detailed record of the production, collective horsepower, and sales of his engines up until the patent expired. He did this for a number of reasons, including tracking those who infringed on the patent rights, maintenance and upgrade details, and especially as data used to extend the patent. With this data, a more clear understanding of the engine's influence is provided. By 1869, nearly 1200 engines had been sold, totaling 118,500 horsepower. Another estimated 60,000 horsepower was being utilized by engines that were created by manufacturers infringing on Corliss's patent, bringing the total horsepower to roughly 180,000. This relatively small amount of engines produced 15% of the United States' total 1.2 million horsepower. The mean horsepower for all Corliss engines in 1870 was 100, while the mean for all steam engines (including Corliss engines) was 30. Some very large engines even allowed for applications as large as 1,400 horsepower. Many were convinced of the Corliss engine's benefits, but adoption was slow due to patent protection. When Corliss was denied a patent extension in 1870, it became a prevalent model for stationary engines in the industrial sector. By the end of the 19th century, the engine was already having a major influence on the manufacturing sector, where it made up only 10% of the sector's engines, but produced 46% of the horsepower. The engine also became a model of efficiency outside of the textile industry as it was used for pumping the waterways of Pawtucket, Rhode Island in 1878 and by playing an essential role in the expansion of the railroad by allowing for very large-scale operations in rolling mills. Many steam engines of the 19th century have been replaced, destroyed, or repurposed, but the longevity of the Corliss engine is apparent today in select distilleries where they are still used as a power source.

Major Applications

Moving from Water to Steam Power

Water power, the world's preceding supply of power, continued to be an essential power source even during the height of steam engine popularity. The steam engine, however, provided many benefits that couldn't be realized by relying solely on water power, allowing it to quickly become industrialised nations' dominant power source (rising from 5% to 80% of the total power in the US from 1838-1860). While many consider the potential for an increase in power generated to be the dominant benefit (with the average horsepower of steam powered mills producing four times the power of water powered mills), others favor the potential for agglomeration. Steam engines made it possible to easily work, produce, market, specialize, viably expand westward without having to worry about the less abundant presence of waterways, and live in communities that weren't geographically isolated in proximity to rivers and streams. Cities and towns were now built around factories where steam engines served as the foundation for the livelihood of many of the citizens. By promoting the agglomeration of individuals, local markets were established and often met with impressive success, cities quickly grew and were eventually urbanized, the quality of living increased as infrastructure was put in place, finer goods could be produced as acquisition of materials became less difficult and expensive, direct local competition led to higher degrees of specialization, and labor and capital were in rich supply. In some counties where the establishments utilized steam power, population growths were even seen to increase. These steam powered towns encouraged growth locally and on the national scale, further validating the economic importance of the steam engine.

The Steamboat

This period of economic growth, which was ushered in by the introduction and adoption of the steamboat, was one of the greatest ever experienced in the United States. Around 1815, steamboats began to replace barges and flatboats in the transport of goods around the United States. Prior to the steamboat, rivers were generally only used in transporting goods from east to west, and from north to south as fighting the current was very difficult and often impossible. Non-powered boats and rafts were assembled up-stream, would carry their cargo down stream, and would often be disassembled at the end of their journey; with their remains being used to construct homes and commercial buildings.

With the steamboat came the need for an improved river system. The natural river system had features that either weren't compatible with steamboat travel, or were only available during certain months when rivers were higher. Some obstacles included rapids, sandbars, shallow waters and waterfalls. To overcome these natural obstacles, a network of canals, locks and dams was constructed. This increased demand for labor spurred tremendous job growth along the rivers.

Following the advent of the steamboat, the United States saw an incredible growth in the transportation of goods and people, which was key in westward expansion. Prior to the steamboat, it could take between three and four months to make the passage from New Orleans to Louisville, averaging twenty miles a day. With the steamboat this time was reduced drastically with trips ranging from twenty-five to thirty-five days. This was especially beneficial to farmers as their crops could now be transported elsewhere to be sold. The steamboat also allowed for increased specialization. Sugar and Cotton were shipped up north while goods like poultry, grain and pork were shipped south. Unfortunately, the steamboat also aided in the internal slave trade.

The economic benefits of the steamboat extended far beyond the construction of the ships themselves, and the goods they transported. These ships led directly to growth in the coal and insurance industries, along with creating demand for repair facilities along the rivers. Additionally the demand for goods in general increased as the steamboat made transport to new destinations both wide reaching and efficient.

Steamboat and Water Transport

After the first steamboat was invented and achieved a number of successful trials, it was quickly adopted and led to an even quicker change in the way of water transport. In 1814, the city of New Orleans recorded 21 steamboat arrivals, but over the course of the following 20 years that number exploded to more than 1200. The steamboat's role as a major transportation source was secured. The transport sector saw enormous growth following the steam engine's application, leading to major innovations in canals, steamboats, and railroads. The steamboat and canal system revolutionized trade of the United States. As the steamboats gained popularity, enthusiasm grew for the building of canals. In 1816, the US had only 100 miles of canals. This needed to change, however, as the potential increase in traded goods from east to west convinced many that canals were a necessary connection between the Mississippi-Ohio waterways with the Great Lakes.

Interchangeable Parts

Interchangeable parts are parts (components) that are, for practical purposes, identical. They are made to specifications that ensure that they are so nearly identical that they will fit into any assembly of the same type. One such part can freely replace another, without any custom fitting (such as filing). This interchangeability allows easy assembly of new devices, and easier repair of existing devices, while minimizing both the time and skill required of the person doing the assembly or repair.

Magneto assembly on the Ford assembly line, 1913. Large crates of identical bolts, nuts and other components can be seen under the conveyor, and the magnetos are themselves also interchangeable.

The concept of interchangeability was crucial to the introduction of the assembly line at the beginning of the 20th century, and has become an important element of some modern manufacturing but is missing from other important industries.

Interchangeability of parts was achieved by combining a number of innovations and improvements in machining operations and the invention of several machine tools, such as the slide rest lathe, screw-cutting lathe, turret lathe, milling machine and metal planer. Additional innovations included jigs for guiding the machine tools, fixtures for holding the workpiece in the proper position, and blocks and gauges to check the accuracy of the finished parts. Electrification allowed individual machine tools to be powered by electric motors, eliminating line shaft drives from steam engines or water power and allowing higher speeds, making modern large scale manufacturing possible. Modern machine tools often have numerical control (NC) which evolved into CNC (computerized numeric control) when microprocessors became available.

Methods for industrial production of interchangeable parts in the United States were first developed in the nineteenth century. The term *American system of manufacturing* was sometimes applied to them at the time, in distinction from earlier methods. Within a few decades such methods were in use in various countries, so *American system* is now a term of historical reference rather than current industrial nomenclature.

First Use

Evidence of the use of interchangeable parts can be traced back over two thousand years to Carthage in the First Punic War. Carthaginian ships had standardized, interchangeable parts that even came with assembly instructions akin to "tab a into slot b" marked on them.

In East Asia during the Warring States period and later the Qin Dynasty, bronze crossbow triggers and locking mechanisms were mass-produced and made to be interchangeable.

Origins of Modern Concept

In the late 18th century, French General Jean-Baptiste Vaquette de Gribeauval promoted standardized weapons in what became known as the *Système Gribeauval* after it was issued as a royal order in 1765. (Its focus at the time was artillery more than muskets or handguns.) One of the accomplishments of the system was that solid cast cannons were bored to precise tolerances, which allowed the walls to be thinner than cannons poured with hollow cores. However, because cores were often off center, the wall thickness determined the size of the bore. Standardized boring allowed cannon to be shorter without sacrificing accuracy and range because of the tighter fit of the shells. It also allowed standardization of the shells.

Before the 18th century, devices such as guns were made one at a time by gunsmiths, and each gun was unique. If one single component of a firearm needed a replacement, the entire firearm either had to be sent to an expert gunsmith for custom repairs, or discarded and replaced by another firearm. During the 18th and early 19th centuries, the idea of replacing these methods with a system of interchangeable manufacture was gradually developed. The development took decades and involved many people.

Gribeauval provided patronage to Honoré Blanc, who attempted to implement the *Système Gribeauval* at the musket level. By around 1778, Honoré Blanc began producing some of the first firearms with interchangeable flint locks, although they were carefully made by craftsmen. Blanc demonstrated in front of a committee of scientists that his muskets could be fitted with flint locks picked at random from a pile of parts.

Muskets with interchangeable locks caught the attention of Thomas Jefferson through the efforts of Honoré Blanc when Jefferson was Ambassador to France in 1785. Jefferson tried to persuade Blanc to move to America, but was not successful, so he wrote to the American Secretary of War with the idea, and when he returned to the USA he worked to fund its development. President George Washington approved of the idea, and by 1798 a contract was issued to Eli Whitney for 12,000 muskets built under the new system.

Louis de Tousard, who fled the French Revolution, joined the U.S. Corp of Artillerists in 1795 and wrote an influential artillerist's manual that stressed the importance of standardization.

Implementation

Numerous inventors began to try to implement the principle Blanc had described. The development of the machine tools and manufacturing practices required would be a great expense to the U.S. Ordnance Department, and for some years while trying to achieve interchangeabililty, the firearms produced cost more to manufacture. By 1853 there was evidence that interchangeable parts, then perfected by the Federal Armories, led to a savings. The Ordnance Department freely shared the techniques used with outside suppliers.

Eli Whitney and An Early Attempt

In the US, Eli Whitney saw the potential benefit of developing "interchangeable parts" for the firearms of the United States military. In July 1801 he built ten guns, all containing the same exact parts and mechanisms, then disassembled them before the United States Congress. He placed the parts in a mixed pile and, with help, reassembled all of the firearms right in front of Congress, much like Blanc had done some years before.

The Congress was captivated and ordered a standard for all United States equipment. Interchangeable parts removed problems concerning the inability to consistently produce new parts for old equipment without significant hand finishing that had plagued the era of unique firearms and equipment. If one firearm part failed, another could be ordered, and the firearm wouldn't have to be discarded. The catch was that Whitney's guns were costly and handmade by skilled workmen.

Whitney was never able to design a manufacturing process capable of producing guns with interchangeable parts. Charles Fitch credited Whitney with successfully executing a firearms contract with interchangeable parts using the American System, but historians Merritt Roe Smith and Robert B. Gordon have since determined that Whitney never achieved interchangeable parts manufacturing. His family's arms company, however, did so after his death.

Brunel's Sailing Blocks

A pulley block for rigging on a sailing ship

Mass production using interchangeable parts was first achieved in 1803 by Marc Is-ambard Brunel in cooperation with Henry Maudslay and Simon Goodrich, under the management of (and with contributions by) Brigadier-General Sir Samuel Bentham, the Inspector General of Naval Works at Portsmouth Block Mills, Portsmouth Dock-yard, Hampshire, England. At the time, the Napoleonic War was at its height, and the Royal Navy was in a state of expansion that required 100,000 pulley blocks to be man-ufactured a year. Bentham had already achieved remarkable efficiency at the docks by introducing power-driven machinery and reorganising the dockyard system.

Marc Brunel, a pioneering engineer, and Maudslay, a founding father of machine tool technology who had developed the first industrially practical screw-cutting lathe in 1800 which standardized screw thread sizes for the first time, collaborated on plans to manufacture block-making machinery; the proposal was submitted to the Admiralty who agreed to commission his services. By 1805, the dockyard had been fully updat-ed with the revolutionary, purpose-built machinery at a time when products were still built individually with different components. A total of 45 machines were required to perform 22 processes on the blocks, which could be made into one of three possible sizes. The machines were almost entirely made of metal thus improving their accuracy and durability. The machines would make markings and indentations on the blocks to ensure alignment throughout the process. One of the many advantages of this new method was the increase in labour productivity due to the less labour-intensive re-quirements of managing the machinery. Richard Beamish, assistant to Brunel's son and engineer, Isambard Kingdom Brunel, wrote:

> "…So that ten men, by the aid of this machinery, can accomplish with uniformi-ty, celerity and ease, what formerly required the uncertain labour of one hun-dred and ten."

By 1808, annual production had reached 130,000 blocks and some of the equipment was still in operation as late as the mid-twentieth century.

Terry's Clocks: Success in Wood

The very first mass production using interchangeable parts in America was, accord-ing to Diana Muir in *Reflections in Bullough's Pond*, Eli Terry's pillar-and-scroll clock, which came off the production line in 1814 at Plymouth, Connecticut. Terry's clocks were made of wooden parts. Making a machine with moving parts mass-produced from metal would be much more difficult.

North and Hall: Success in Metal

The crucial step toward interchangeability in metal parts was taken by Simeon North, working only a few miles from Eli Terry. North created one of the world's first true milling machines to do metal shaping that had been done by hand with a file. Diana

Muir believes that North's milling machine was online around 1816. Muir, Merritt Roe Smith, and Robert B. Gordon all agree that before 1832 both Simeon North and John Hall were able to mass-produce complex machines with moving parts (guns) using a system that entailed the use of rough-forged parts, with a milling machine that milled the parts to near-correct size, and that were then "filed to gage by hand with the aid of filing jigs."

Historians differ over the question of whether Hall or North made the crucial improvement. Merrit Roe Smith believes that it was done by Hall. Muir demonstrates the close personal ties and professional alliances between Simeon North and neighboring mechanics mass-producing wooden clocks to argue that the process for manufacturing guns with interchangeable parts was most probably devised by North in emulation of the successful methods used in mass-producing clocks. It may not be possible to resolve the question with absolute certainty unless documents now unknown should surface in the future.

Late 19th and Early 20th Centuries: Dissemination Throughout Manufacturing

Skilled engineers and machinists, many with armory experience, spread interchangeable manufacturing techniques to other American industries including clockmakers and sewing machine manufacturers Wilcox and Gibbs and Wheeler and Wilson, who used interchangeable parts before 1860. Late to adopt the interchangeable system were Singer Corporation sewing machine (1870s), reaper manufacturer McCormick Harvesting Machine Company (1870s-80s) and several large steam engine manufacturers such as Corliss (mid-1880s) as well as locomotive makers. Typewriters followed some years later by large scale of production of bicycles in the 1880s used the interchangeable system.

During these decades, true interchangeability grew from a scarce and difficult achievement into an everyday capability throughout the manufacturing industries. In the 1950s and 1960s, historians of technology broadened the world's understanding of the history of the development. Few people outside that academic discipline knew much about the topic until as recently as the 1980s and 1990s, when the academic knowledge began finding wider audiences. As recently as the 1960s, when Alfred P. Sloan published his famous memoir and management treatise, *My Years with General Motors*, even the longtime president and chair of the largest manufacturing enterprise that had ever existed knew very little about the history of the development, other than to say that "[Henry M. Leland was], I believe, one of those mainly responsible for bringing the technique of interchangeable parts into automobile manufacturing. [...] It has been called to my attention that Eli Whitney, long before, had started the development of interchangeable parts in connection with the manufacture of guns, a fact which suggests a line of descent from Whitney to Leland to the automobile industry." One of the better-known books on the subject, which was first published in 1984 and has enjoyed a readership beyond academia, has been David A. Hounshell's *From the American System to Mass Production, 1800-1932: The Development of Manufacturing Technology in the United States*.

Socioeconomic Context

The principle of interchangeable parts flourished and developed throughout the 19th century, and led to mass production in many industries. It was based on the use of templates and other jigs and fixtures, applied by semi-skilled labor using machine tools to augment (and later largely replace) the traditional hand tools. Throughout this century there was much development work to be done in creating gauges, measuring tools (such as calipers and micrometers), standards (such as those for screw threads), and processes (such as scientific management), but the principle of interchangeability remained constant. With the introduction of the assembly line at the beginning of the 20th century, interchangeable parts became ubiquitous elements of manufacturing.

Selective Assembly

Interchangeability relies on parts' dimensions falling within the tolerance range. The most common mode of assembly is to design and manufacture such that, as long as each part that reaches assembly is within tolerance, the mating of parts can be totally random. This has value for all the reasons already discussed earlier.

There is another mode of assembly, called "selective assembly", which gives up *some* of the randomness capability in trade-off for other value. There are two main areas of application that benefit economically from selective assembly: when tolerance ranges are so tight that they cannot quite be held reliably (making the total randomness unavailable); and when tolerance ranges can be reliably held, but the fit and finish of the final assembly is being maximized by voluntarily giving up some of the randomness (which makes it available but not ideally desirable). In either case the principle of selective assembly is the same: parts are *selected* for mating, rather than being mated at random. As the parts are inspected, they are graded out into separate bins based on what end of the range they fall in (or violate). Falling within the high or low end of a range is usually called being *heavy* or *light*; violating the high or low end of a range is usually called being *oversize* or *undersize*. Examples are given below.

French and Vierck provide a one-paragraph description of selective assembly that aptly summarizes the concept.

One might ask, if parts must be selected for mating, then what makes selective assembly any different from the oldest craft methods? But there is in fact a significant difference. Selective assembly merely grades the parts into several *ranges*; within each range, there is still random interchangeability. This is quite different from the older method of fitting by a craftsman, where each mated set of parts is specifically filed to fit each part with a *specific, unique* counterpart.

Random Assembly not Available: Oversize and Undersize Parts

In contexts where the application requires extremely tight (narrow) tolerance rang
es, the requirement may push slightly past the limit of the ability of the machining
and other processes (stamping, rolling, bending, etc.) to stay within the range. In
such cases, selective assembly is used to compensate for a lack of *total* interchangeability among the parts. Thus, for a pin that must have a sliding fit in its hole (free
but not sloppy), the dimension may be spec'd as 12.00 +0 -0.01 mm for the pin, and
12.00 +.01 -0 for the hole. Pins that came out oversize (say a pin at 12.003mm diameter) are not necessarily scrap, *but* they can only be mated with counterparts that
also came out oversize (say a hole at 12.013mm). The same is then true for matching
*under*size parts with *under*size counterparts. Inherent in this example is that for this
product's application, the 12 mm dimension does not require extreme *accuracy*, *but*
the desired fit between the parts does require good *precision*. This allows the makers
to "cheat a little" on *total* interchangeability in order to get more value out of the
manufacturing effort by reducing the rejection rate (scrap rate). This is a sound engineering decision as long as the application and context support it. For example, for
machines for which there is no intention for any future field service of a parts-replac-
ing nature (but rather only simple replacement of the whole unit), this makes good
economic sense. It lowers the unit cost of the products, and it doesn't impede future
service work.

An example of a product that might benefit from this approach could be a car transmission where there is no expectation that the field service person will repair the
old transmission; instead, he will simply swap in a new one. Therefore, total interchangeability was not absolutely required for the assemblies inside the transmissions. It would have been specified anyway, simply on general principle, except for
a certain shaft that required precision so high as to cause great annoyance and high
scrap rates in the grinding area, but for which only decent accuracy was required, as
long as the fit with its hole was good in every case. Money could be saved by saving
many shafts from the scrap bin.

Economic and Commercial Realities

Examples like the one above are not as common in real commerce as they con-
ceivably could be, mostly because of separation of concerns, where each part of
a complex system is expected to give performance that does not make any limiting assumptions about other parts of the system. In the car transmission example,
the separation of concerns is that individual firms and customers accept no lack of
freedom or options from others in the supply chain. For example, in the car buyer's
view, the car manufacturer is "not within its rights" to assume that no field-service
mechanic will ever repair the old transmission instead of replacing it. The customer
expects that that decision will be preserved for *him* to make later, at the repair shop,
based on which option is less expensive for him at that time (figuring that replacing

one shaft is cheaper than replacing a whole transmission). This logic is not always valid in reality; it might have been better for the customer's total ownership cost to pay a lower initial price for the car (especially if the transmission service is covered under the standard warranty for 10 years, and the buyer intends to replace the car before then anyway) than to pay a higher initial price for the car but preserve the option of total interchangeability of every last nut, bolt, and shaft throughout the car (when it is not going to be taken advantage of anyway). But commerce is generally too chaotically multivariate for this logic to prevail, so total interchangeability ends up being specified and achieved even when it adds expense that was "needless" from a holistic view of the commercial system. But this may be avoided to the extent that customers experience the *overall* value (which their minds can detect and appreciate) without having to understand its logical analysis. Thus buyers of an amazingly affordable car (surprisingly low initial price) will probably never complain that the transmission was not field-serviceable as long as *they themselves* never had to pay for transmission service in the lifespan of their ownership. This analysis can be important for the manufacturer to understand (even if it is lost on the customer), because he can carve for himself a competitive advantage in the marketplace if he can accurately predict where to "cut corners" in ways that the customer will never have to pay for. Thus he could give himself lower transmission unit cost. However, he must be sure when he does so that the transmissions he's using are reliable, because their replacement, being covered under a long warranty, will be at his expense.

Random Assembly Available But Not Ideally Desirable: "Light" and "Heavy" Parts

The other main area of application for selective assembly is in contexts where total interchangeability is in fact achieved, but the "fit and finish" of the final products can be enhanced by minimizing the dimensional mismatch between mating parts. Consider another application similar to the one above with the 12 mm pin. But say that in this example, not only is the *precision* important (to produce the desired fit), but the *accuracy* is also important (because the 12 mm pin must interact with something else that will have to be accurately sized at 12 mm). Some of the implications of this example are that the rejection rate cannot be lowered; all parts must fall within tolerance range or be scrapped. So there are no savings to be had from salvaging oversize or undersize parts from scrap, then. However, there *is* still one bit of value to be had from selective assembly: having all the mated pairs have as close to identical sliding fit as possible (as opposed to some tighter fits and some looser fits—all sliding, but with varying resistance).

An example of a product that might benefit from this approach could be a toolroom-grade machine tool, where not only is the accuracy highly important but also the fit and finish.

Paper Machine

Many modern papermaking machines are based on the principles of the Fourdrinier Machine, which uses a specially woven plastic fabric mesh conveyor belt (known as a *wire* as it was once woven from bronze) in the forming section, where a slurry of fibre (usually wood or other vegetable fibres) is drained to create a continuous paper web. After the forming section the wet web passes through a press section to squeeze out excess water, then the pressed web passes through a heated drying section.

The original Fourdrinier forming section used a horizontal drainage area, referred to as the *drainage table*.

Paper machines have four distinct operational sections:

- Forming section, commonly called the wet end, is where the slurry of fibres filters out fluid a continuous fabric loop to form a wet web of fibre.

- Press section where the wet fibre web passes between large rolls loaded under high pressure to squeeze out as much water as possible.

- Drying section, where the pressed sheet passes partly around, in a serpentine manner, a series of steam heated drying cylinders. Drying removes the water content down to a level of about 6%, where it will remain at typical indoor atmospheric conditions.

- Calender section where the dried paper is smoothened under high loading and pressure. Only one *nip* (where the sheet is pressed between two rolls) is necessary in order to hold the sheet, which shrinks through the drying section and is held in tension between the press section (or breaker stack if used) and the calender. Extra nips give more smoothing but at some expense to paper strength.

Paper machines are long-lived assets that usually remain in service for several decades. It is common to rebuild machines periodically to increase production and improve quality or to change the paper grade.

History of Paper Machines

Before the invention of continuous paper making, paper was made in individual sheets by stirring a container of pulp slurry and either pouring it into a fabric sieve called a sheet mould or dipping and lifting the sheet mould from the vat. While still on the fabric in the sheet mould the wet paper is pressed to remove excess water and then the sheet was lifted off to be hung over a rope or wooden rod to air dry. In 1799, Louis-Nicolas Robert of Essonnes, France, was granted a patent for a continuous paper making machine. At the time Robert was working for Saint-Léger Didot, with whom he quarrelled over the ownership of the invention. Didot thought that England was a better place to develop the machine. But during the troubled times of the French Revolution, he could not go there himself, so he sent his brother in law, John Gamble, an Englishman living in Paris. Through a chain of acquaintances, Gamble was introduced to the brothers Sealy and Henry Fourdrinier, stationers of London, who agreed to finance the project. Gamble was granted British patent 2487 on 20 October 1801.

With the help particularly of Bryan Donkin, a skilled and ingenious mechanic, an improved version of the Robert original was installed at Frogmore Mill, Apsley, Hertfordshire, in 1803, followed by another in 1804. A third machine was installed at the Fourdriniers' own mill at Two Waters. The Fourdriniers also bought a mill at St Neots intending to install two machines there and the process and machines continued to develop.

Thomas Gilpin is most often credited for creating the first U.S cylinder type papermaking machine at Brandywine Creek, Delaware in 1817. This machine was also developed in England, but it was a cylinder mould machine. The Fourdrinier machine wasn't introduced into the USA until 1827.

However, records show Charles Kinsey of Patterson, NJ had already patented a continuous process papermaking machine in 1807. Kinsey's machine was built locally by Daniel Sawn and by 1809 the Kinsey machine was successfully making paper at the Essex Mill in Paterson. Financial stress and potential opportunities created by the Embargo of 1807 eventually persuaded Kinsey and his backers to change the mill's focus from paper to cotton and Kinsey's early papermaking successes were soon overlooked and forgotten.

Gilpin's 1817 patent was similar to Kinsey's, as was the John Ames patent of 1822. The Ames patent was challenged by his competitors, asserting that Kinsey was the original inventor and Ames had been pilfering other peoples' ideas, their evidence being the employment of Daniel Sawn to work on his machine.

The method of continuous production demonstrated by the paper machine influenced the development of continuous rolling of iron and later steel and other continuous production processes.

Pulp Types and Their Preparations

The plant fibres used for pulp are composed mostly of cellulose and hemi-cellulose, which have a tendency to form molecular linkages between fibres in the presence of water. After the water evaporates the fibres remain bonded. It is not necessary to add additional binders for most paper grades, although both wet and dry strength additives may be added.

Rags of cotton and linen were the major source of pulp for paper before wood pulp. Today almost all pulp is of wood fibre. Cotton fibre is used in speciality grades, usually in printing paper for such things as resumes and currency.

Sources of rags often appear as waste from other manufacturing such as denim fragments or glove cuts. Fibres from clothing come from the cotton boll. The fibres can range from 3 to 7 cm in length as they exist in the cotton field. Bleach and other chemicals remove the colour from the fabric in a process of cooking, usually with steam. The cloth fragments mechanically abrade into fibres, and the fibres get shortened to a length appropriate for manufacturing paper with a cutting process. Rags and water dump into a trough forming a closed loop. A cylinder with cutting edges, or knives, and a knife bed is part of the loop. The spinning cylinder pushes the contents of the trough around repeatedly. As it lowers slowly over a period of hours, it breaks the rags up into fibres, and cuts the fibres to the desired length. The cutting process terminates when the mix has passed the cylinder enough times at the programmed final clearance of the knives and bed.

Another source of cotton fibre comes from the cotton ginning process. The seeds remain, surrounded by short fibres known as linters for their short length and resemblance to lint. Linters are too short for successful use in fabric. Linters removed from the cotton seeds are available as first and second cuts. The first cuts are longer.

The two major classifications of pulp are chemical and mechanical. Chemical pulps formerly used a sulphite process, but the kraft process is now predominant. Kraft pulp has superior strength to sulphite and mechanical pulps. Both chemical pulps and mechanical pulps may be bleached to a high brightness.

Chemical pulping dissolves the lignin that bonds fibres to one another, and binds the outer fibrils that compose individual fibres to the fibre core. Lignin, like most other substances that can separate fibres from one another, acts as a debonding agent, lowering strength. Strength also depends on maintaining long cellulose molecule chains. The kraft process, due to the alkali and sulphur compounds used, tends to minimize attack on the cellulose and the non-crystalline hemicellulose, which promotes bonding, while dissolving the lignin. Acidic pulping processes shorten the cellulose chains.

Kraft pulp makes superior linerboard and excellent printing and writing papers.

Groundwood, the main ingredient used in newsprint and a principal component of magazine papers (coated publications), is literally ground wood produced by a grinder. Therefore, it contains a lot of lignin, which lowers its strength. The grinding produces very short fibres that drain slowly.

Thermomechanical pulp (TMP) is a variation of groundwood where fibres are separated mechanically while at high enough temperatures to soften the lignin.

Between chemical and mechanical pulps there are semi-chemical pulps that use a mild chemical treatment followed by refining. Semi-chemical pulp is often used for corrugating medium.

Bales of recycled paper (normally old corrugated containers) for unbleached (brown) packaging grades may be simply pulped, screened and cleaned. Recycling to make white papers is usually done in a deinking plant, which employs screening, cleaning, washing, bleaching and flotation. Deinked pulp is used in printing and writing papers and in tissue, napkins and paper towels. It is often blended with virgin pulp.

At integrated pulp and paper mills, pulp is usually stored in high density towers before being pumped to stock preparation. Non integrated mills use either dry pulp or wet lap (pressed) pulp, usually received in bales. The pulp bales are slushed in a [re]pulper.

Stock (Pulp) Preparation

Stock preparation is the area where pulp is usually refined, blended to the appropriate proportion of hardwood, softwood or recycled fibre, and diluted to as uniform and constant as possible consistency. The pH is controlled and various fillers, such as whitening agents, size and wet strength or dry strength are added if necessary. Additional fillers such as clay, calcium carbonate and titanium dioxide increase opacity so printing on reverse side of a sheet will not distract from content on the obverse side of the sheet. Fillers also improve printing quality.

Pulp is pumped through a sequence of tanks that are commonly called *chests*, which may be either round or more commonly rectangular. Historically these were made of special ceramic tile faced reinforced concrete, but mild and stainless steels are also used. Low consistency pulp slurries are kept agitated in these chests by propeller like agitators near the pump suction at the chest bottom.

In the following process, different types of pulp, if used, are normally treated in separate but similar process lines until combined at a blend chest:

From high density storage or from slusher/pulper the pulp is pumped to a low density storage chest (tank). From there it is typically diluted to about 4% consistency before being pumped to an unrefined stock chest. From the unrefined stock chest stock is again pumped, with consistency control, through a refiner. Refining is an operation whereby the pulp slurry passes between a pair of discs, one of which is stationary and the other rotating at speeds of typically 1,000 or 1,200 RPM for 50 and 60 Hz AC, respectively. The discs have raised bars on their faces and pass each other with narrow clearance. This action unravels the outer layer of the fibres, causing the fibrils of the fibres to partially detach and bloom outward, increasing the surface area to promoting bonding. Refining thus increases tensile strength. For example, tissue paper is relatively unrefined whereas packaging paper is more highly refined. Refined stock from the refiner then goes to a refined stock chest, or blend chest, if used as such.

Hardwood fibres are typically 1 mm long and smaller in diameter than the 4 mm length typical of softwood fibres. Refining can cause the softwood fibre tube to collapse resulting in undesirable properties in the sheet.

From the refined stock, or blend chest, stock is again consistency controlled as it is being pumped to a machine chest. It may be refined or additives may be added en route to the machine chest.

The machine chest is basically a consistency levelling chest having about 15 minutes retention. This is enough retention time to allow any variations in consistency entering the chest to be levelled out by the action of the basis weight valve receiving feedback from the on line basis weight measuring scanner. (Note: Many paper machines mistakenly control consistency coming out of the machine chest, interfering with basis weight control.)

Sections of The Paper Machine

There are four main sections on this paper machine. In practice calender rolls are normally placed vertically in a *stack*.

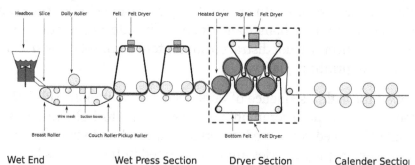

| Wet End | Wet Press Section | Dryer Section | Calender Section |

Diagram showing the sections of the Fourdrinier machine

Forming Section or Wet End

A worker inspecting wet, bleached wood pulp

From the machine chest stock is pumped to a head tank, commonly called a *head box*, whose purpose is to maintain a constant head (pressure) on the stock as it feeds the basis weight valve. The head box also provides a means allowing air bubbles to escape. The basis weight valve meters the stock to the recirculating stream of water that is pumped, by the *fan pump*, from a whitewater chest through to the flow box. On the way to the flow box the pulp slurry may pass through centrifugal cleaners, which remove heavy contaminants like sand, and screens, which break up fibre clumps and remove over-sized debris.

Wood fibers have a tendency to attract one another, forming clumps, the effect being called flocculation. Flocculation is lessened by lowering consistency and or by agitating the slurry; however, de-flocculation becomes very difficult at much above 0.5% consistency. Minimizing the degree of flocculation when forming is important to physical properties of paper.

The consistency in the flow box is typically under 0.4% for most paper grades, with longer fibres requiring lower consistency than short fibres. Higher consistency causes more fibres to be oriented in the z direction, while lower consistency promotes fibre orientation in the x-y direction. Higher consistency promotes higher calliper (thickness) and stiffness, lower consistency promotes higher tensile and some other strength properties and also improves formation (uniformity). Many sheet properties continue to improve down to below 0.1% consistency; however, this is an impractical amount of water to handle. (Most paper machine run a higher headbox consistency than optimum because they have been sped up over time without replacing the fan pump and headbox. There is also an economic trade off with high pumping costs for lower consistency).

The stock slurry, often called *white water* at this point, exits the flow box through a rectangular opening of adjustable height called the *slice*, the white water stream being called the *jet* and it is pressurized on high speed machines so as to land gently on the moving fabric loop or *wire* at a speed typically between plus or minus 3% of the wire speed, called *rush* and *drag* respectively. Excessive *rush* or *drag* causes more orientation of fibres in the machine direction and gives differing physical properties in machine and cross directions; however, this phenomenon is not completely avoidable on Fourdrinier machines.

On lower speed machines at 700 feet per minute, gravity and the height of the stock in the headbox creates sufficient pressure to form the jet through the opening of the slice. The height of the stock is the head, which gives the headbox its name. The speed of the jet compared to the speed of the wire is known as the *jet-to-wire ratio*. When the jet-to-wire ratio is less than unity, the fibres in the stock become drawn out in the machine direction. On slower machines where sufficient liquid remains in the stock before draining out, the wire can be driven back and forth with a process known as *shake*. This provides some measure of randomizing the direction of the fibres and gives the sheet more uniform strength in both the machine and cross-machine directions. On fast machines, the stock does not remain on the wire in liquid form long enough and the long fibres line up with the machine. When the jet-to-wire ratio exceeds unity, the fibers tend to pile up in lumps. The resulting variation in paper density provides the antique or parchment paper look.

Two large rolls typically form the ends of the drainage section, which is called the *drainage table*. The *breast roll* is located under the flow box, the jet being aimed to land on it at about the top centre. At the other end of the drainage table is the suction (*couch*) roll. The couch roll is a hollow shell, drilled with many thousands of precisely spaced holes of about 4 to 5 mm diameter. The hollow shell roll rotates over a stationary suction box, normally placed at the top centre or rotated just down machine. Vacuum is pulled on the suction box, which draws water from the web into the suction box. From the suction roll the sheet feeds into the press section.

Down machine from the suction roll, and at a lower elevation, is the *wire turning roll*. This roll is driven and pulls the wire around the loop. The wire turning roll has a considerable angle of wrap in order to grip the wire.

Supporting the wire in the drainage table area are a number of drainage elements. In addition to supporting the wire and promoting drainage, the elements de-flocculate the sheet. On low speed machines these table elements are primarily *table rolls*. As speed increases the suction developed in the nip of a table roll increases and at high enough speed the wire snaps back after leaving the vacuum area and causes stock to jump off the wire, disrupting the formation. To prevent this drainage foils are used. The foils are typically sloped between zero and two or three degrees and give a more gentle action. Where rolls and foils are used, rolls are used near the headbox and foils further down machine.

Approaching the dry line on the table are located low vacuum boxes that are drained by a barometric leg under gravity pressure. After the dry line are the suction boxes with applied vacuum. Suction boxes extend up to the couch roll. At the couch the sheet consistency should be about 25%.

Variations of The Fourdrinier Forming Section

The forming section type is usually based on the grade of paper or paperboard being produced; however, many older machines use a less than optimum design. Older machines can be upgraded to include more appropriate forming sections.

A second headbox may be added to a conventional fourdrinier to put a different fibre blend on top of a base layer. A *secondary headbox* is normally located at a point where the base sheet is completely drained. This is not considered a separate ply because the water action does a good job of intermixing the fibers of the top and bottom layer. Secondary headboxes are common on linerboard.

A modification to the basic fourdrinier table by adding a second wire on top of the drainage table is known as a top wire former. The bottom and top wires converge and some drainage is up through the top wire. A top wire improves formation and also gives more drainage, which is useful for machines that have been sped up.

The Twin Wire Machine or Gap former uses two vertical wires in the forming section, thereby increasing the de-watering rate of the fibre slurry while also giving uniform two sidedness.

There are also machines with entire Fourdrinier sections mounted above a traditional Fourdrinier. This allows making multi-layer paper with special characteristics. These are called top Fourdriniers and they make multi-ply paper or paperboard. Commonly this is used for making a top layer of bleached fibre to go over an unbleached layer.

Another type forming section is the cylinder mould machine using a mesh-covered rotating cylinder partially immersed in a tank of fibre slurry in the wet end to form a paper web, giving a more random distribution of the cellulose fibres. Cylinder machines can form a sheet at higher consistency, which gives a more three dimensional fibre orientation than lower consistencies, resulting in higher calliper (thickness) and more stiffness in the machine direction (MD). High MD stiffness is useful in food packaging like cereal boxes and other boxes like dry laundry detergent.

Tissue machines typically form the paper web between a wire and a special fabric (felt) as they wrap around a forming roll. The web is pressed from the felt directly onto a large diameter dryer called a *yankee*. The paper sticks to the yankee dryer and is peeled off with a scraping blade called a *doctor*. Tissue machines operate at speeds of up to 2000 m/min.

Press Section

Granite press roll at a granite quarry site   Paper machine

The second section of the paper machine is the press section, which removes much of the remaining water via a system of nips formed by rolls pressing against each other aided by press felts that support the sheet and absorb the pressed water. The paper web consistency leaving the press section can be above 40%.

Pressing is the most efficient method of de-watering the sheet as only mechanical action is required. Press felts historically were made from wool. However, today they are nearly 100% synthetic. They are made up of a polyamide woven fabric with thick batt applied in a specific design to maximise water absorption.

Presses can be single or double felted. A single felted press has a felt on one side and a smooth roll on the other. A double felted press has both sides of the sheet in contact with a press felt. Single felted nips are useful when mated against a smooth roll (usually in the top position), which adds a two-sidedness—making the top side appear smoother than the bottom. Double felted nips impart roughness on both sides of the sheet. Double felted presses are desirable for the first press section of heavy paperboard.

Simple press rolls can be rolls with grooved or blind drilled surface. More advanced press rolls are suction rolls. These are rolls with perforated shell and cover. The shell made of metal material such as bronze stainless steel is covered with rubber or a synthetic material. Both shell and cover are drilled throughout the surface. A stationary suction box is fitted in the core of the suction roll to support the shell being pressed. End face mechanical seals are used for the interface between the inside surface of the shell and the suction box. For the smooth rolls, they are typically made of granite rolls. The granite rolls can be up to 30-foot (9.1 m) long and 6 feet (1.8 m) in diameter.

Conventional roll presses are configured with one of the press rolls is in a fixed position, with a mating roll being loaded against this fixed roll. The felts run through the nips of the press rolls and continues around a felt run, normally consisting of several felt rolls. During the dwell time in the nip, the moisture from the sheet is transferred to the press felt. When the press felt exits the nip and continues around, a vacuum box known as an Uhle Box applies vacuum (normally -60 kPa) to the press felt to remove the moisture so that when the felt returns to the nip on the next cycle, it does not add moisture to the sheet.

Some grades of paper use suction pick up rolls that use vacuum to transfer the sheet from the couch to a lead in felt on the first press or between press sections. Pickup roll presses normally have a vacuum box that has two vacuum zones (low vacuum and high vacuum). These rolls have a large number of drilled holes in the cover to allow the vacuum to pass from the stationary vacuum box through the rotating roll covering. The low vacuum zone picks up the sheet and transfers, while the high vacuum zone attempts to remove moisture. Unfortunately, at high enough speed centrifugal force flings out vacuumed water, making this less effective for dewatering. Pickup presses also have standard felt runs with Uhle boxes. However, pickup press design is quite different, as air movement is important for the pickup and dewatering facets of its role.

Crown Controlled Rolls (also known as CC Rolls) are usually the mating roll in a press arrangement. They have hydraulic cylinders in the press rolls that ensure that the roll does not bow. The cylinders connect to a shoe or multiple shoes to keep the crown on the roll flat, to counteract the natural "bend" in the roll shape due to applying load to the edges.

Extended Nip Presses (or ENP) are a relatively modern alternative to conventional roll presses. The top roll is usually a standard roll, while the bottom roll is actually a large CC roll with an extended shoe curved to the shape of the top roll, surrounded by a rotating rubber belt rather than a standard roll cover. The goal of the ENP is to extend the dwell time of the sheet between the two rolls thereby maximising the de-watering. Compared to a standard roll press that achieves up to 35% solids after pressing, an ENP brings this up to 45% and higher—delivering significant steam savings or speed increases. ENPs densify the sheet, thus increasing tensile strength and some other physical properties.

Dryer Section

Dryer section of an older Fourdrinier-style paper-making machine. These narrow, small diameter dryers are not enclosed by a hood, dating the photo to before the 1970s.

The dryer section of the paper machine, as its name suggests, dries the paper by way of a series of internally steam-heated cylinders that evaporate the moisture. Steam pressures may range up to 160 psig. Steam enters the end of the dryer head (cylinder cap) through a steam joint and condensate exits through a siphon that goes from the internal shell to a centre pipe. From the centre pipe the condensate exits through a joint on the dryer head. Wide machines require multiple siphons. In fast machines centrifugal force holds the condensate layer still against the shell and turbulence generating bars are typically used to agitate the condensate layer and improve heat transfer.

The sheet is usually held against the dryers by long felt loops on the top and bottom of each dryer section. The felts greatly improve heat transfer. Dryer felts are made of coarse thread and have a very open weave that is almost see through, It is common to have the first bottom dryer section unfelted to dump broke on the basement floor during sheet breaks or when threading the sheet.

Paper dryers are typically arranged in groups called *sections* so that they can be run at a progressively slightly slower speed to compensate for sheet shrinkage as the paper dries. The gaps between sections are called *draws*.

The drying sections are usually enclosed to conserve heat. Heated air is usually supplied to the pockets where the sheet breaks contact with the driers. This increases the rate of drying. The pocket ventilating tubes have slots along their entire length that face into the pocket. The dryer hoods are usually exhausted with a series of roof mounted hood exhausts fans down the dryer section.

SC Sizer

Additional sizing agents, including resins, glue, or starch, can be added to the web to alter its characteristics. Sizing improves the paper's water resistance, decreases its ability to fuzz, reduces abrasiveness, and improves its printing properties and surface bond strength. These may be applied at the wet (internal sizing) or on the dry end (surface sizing), or both. At the dry end sizing is usually applied with a *size press*. The size press may be a roll applicator (flooded nip) or Nozzle applicator . It is usually placed before the last dryer section. Some paper machines also make use of a 'coater' to apply a coating of fillers such as calcium carbonate or china clay usually suspended in a binder of cooked starch and styrene-butadiene latex. Coating produces a very smooth, bright surface with the highest printing qualities.

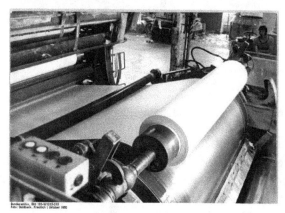

Paper leaving the machine is rolled onto a *reel* for further processing.

Calender Section

A calender consists of two or more rolls, where pressure is applied to the passing paper. Calenders are used to make the paper surface extra smooth and glossy. It also gives it a more uniform thickness. The pressure applied to the web by the rollers determines the finish of the paper.

After calendering, the web has a moisture content of about 6% (depending on the furnish). It is wound onto a roll called a *tambour* or *reel*, and stored for final cutting and shipping. The roll hardness should be checked, obtained and adjusted accordingly to insure that the roll hardness is within the acceptable range for the product.

Glossary

broke: waste paper, either made during a sheet break or trimmings. It is gathered up and put in a repulper for recycling back into the process.

consistency: the percent dry fibre in a pulp slurry.

couch: French meaning *to lie down*. Following the couch roll the sheet is lifted off the wire and transferred into the press section.

dandy roll: a mesh covered hollow roll that rides on top of the Fourdrinier. It breaks up fibre clumps to improve the sheet formation and can also be used to make an imprint, as with laid paper.

fan pump: the large pump that circulates white water from the white water chest to the headbox. The flow may go through screens and cleaners, if used. On large paper machines fan pumps may be rated in tens of thousands of gallons per minute.

felt: a loop of fabric or synthetic material that goes between press rolls and serves as a place to receive the pressed out water. Felts also support the wet paper web and guide it through the press section. Felts are also used in the dryer section to keep the sheet in close contact with the dryers and increase heat transfer.

filler: a finely divided substance added to paper in the forming process. Fillers improve print quality, brightness and opacity. The most common fillers are clay and calcium carbonate. Titanium dioxide is a filler but also improves brightness and opacity. Use of calcium carbonate filler is the process called *alkaline sizing* and uses different chemistry than acid sizing. Alkaline sized paper has superior ageing properties.

formation: the degree of uniformity of fiber distribution in finished paper, which is easily seen by holding paper up to the light.

headbox: the pressure chamber where turbulence is applied to break up fibre clumps in the slurry. The main job of the headbox is to distribute the fiber slurry uniformly across the wire.

nip: the contact area where two opposing rolls meet, such as in a press or calender

pH: the degree of acidity or alkalinity of a solution. Alkaline paper has a very long life. Acid paper deteriorates over time, which caused libraries to either take conservation measures or replace many older books.

size: a chemical (formerly rosin derived but now a different chemical) or starch, applied to paper to retard the rate of water penetration. Sizing prevents *bleeding* of ink during printing, improving the sharpness of printing.

slice: the adjustable rectangular orifice, usually at the bottom of the headbox, through which the whitewater jet discharges onto the wire. The slice opening and water pressure together determine the amount and velocity of whitewater flow through the slice. The slice usually has some form of adjustment mechanism to even out the paper weight profile across the machine (CD profile), although a newer methods is to inject water into the whitewater across the headbox slice area, thereby using localized consistency to control CD weight profile.

stock: a pulp slurry that has been processed in the stock preparation area with necessary additives, refining and pH adjustment and ready for making paper

web: the continuous flow of un-dried fibre from the couch roll down the paper machine

white water: filtrate from the drainage table. The white water from the table is usually stored in a white water chest from which it is pumped by the fan pump to the headbox.

wire: the woven mesh fabric loop that is used for draining the pulp slurry from the headbox. Until the 1970s bronze wires were used but now they are woven from coarse mono-filament synthetics similar to fishing line but very stiff.

Coal Mining

The goal of coal mining is to obtain coal from the ground. Coal is valued for its energy content, and, since the 1880s, has been widely used to generate electricity. Steel and cement industries use coal as a fuel for extraction of iron from iron ore and for cement production. In the United Kingdom and South Africa, a coal mine and its structures are a colliery. In Australia, "colliery" generally refers to an underground coal mine. In the United States "colliery" has historically been used to describe a coal mine operation, but the word today is not commonly used.

A painting depicting men leaving a UK colliery at the close of a shift.

Surface coal mining in Wyoming in the United States.

A coal mine in Bihar, India.

A coal mine in Frameries, Belgium.

Coal mining has had many developments over the recent years, from the early days of men tunneling, digging and manually extracting the coal on carts, to large open cut and long wall mines. Mining at this scale requires the use of draglines, trucks, conveyors, jacks and shearers

History

Ships were used to haul coal.

Small-scale mining of surface deposits dates back thousands of years. For example, in Roman Britain, the Romans were exploiting all major coalfields (save those of North and South Staffordshire) by the late 2nd century AD. While much of its use remained local, a lively trade developed along the North Sea coast supplying coal to Yorkshire and London.

The Industrial Revolution, which began in Britain in the 18th century, and later spread to continental Europe and North America, was based on the availability of coal to power steam engines. International trade expanded exponentially when coal-fed steam engines were built for the railways and steamships. The new mines that grew up in the 19th century depended on men and children to work long hours in often dangerous working conditions. There were many coalfields, but the oldest were in Newcastle and Durham, South Wales, the Central Belt of Scotland and the Midlands, such as those at Coalbrookdale.

Children would usually work as trappers; this is where they had to open and close trap doors to allow mine carts in. This made sure no harmful gases built up in the mine.

The oldest continuously worked deep-mine in the United Kingdom is Tower Colliery in South Wales valleys in the heart of the South Wales coalfield. This colliery was developed in 1805, and its miners bought it out at the end of the 20th century, to prevent it from being closed. Tower Colliery was finally closed on 25 January 2008, although production continued at the Aberpergwm drift mine owned by Walter Energy until suspended on 3 July 2015.

Coal was mined in America in the early 18th century, and commercial mining started around 1730 in Midlothian, Virginia.

Coal-cutting machines were invented in the 1880s. Before this invention, coal was mined from underground with a pick and shovel. By 1912, surface mining was conducted with steam shovels designed for coal mining.

Methods of Extraction

The most economical method of coal extraction from coal seams depends on the depth and quality of the seams, and the geology and environmental factors. Coal mining processes are differentiated by whether they operate on the surface or underground. Many coals extracted from both surface and underground mines require washing in a coal preparation plant. Technical and economic feasibility are evaluated based on the following: regional geological conditions; overburden characteristics; coal seam continuity, thickness, structure, quality, and depth; strength of materials above and below the seam for roof and floor conditions; topography (especially altitude and slope); climate; land ownership as it affects the availability of land for mining and access; surface drainage patterns; ground water conditions; availability of labor and materials; coal purchaser requirements in terms of tonnage, quality, and destination; and capital investment requirements.

Surface mining and deep underground mining are the two basic methods of mining. The choice of mining method depends primarily on depth of burial, density of the over-burden and thickness of the coal seam. Seams relatively close to the surface, at depths less than approximately 180 ft (50 m), are usually surface mined.

Coal that occurs at depths of 180 to 300 ft (50 to 100 m) are usually deep mined, but in some cases surface mining techniques can be used. For example, some western U.S. coal that occur at depths in excess of 200 ft (60 m) are mined by the open pit methods, due to thickness of the seam 60–90 feet (20–30 m). Coals occurring below 300 ft (100 m) are usually deep mined. However, there are open pit mining operations working on coal seams up to 1000–1500 feet (300–450 m) below ground level, for instance Tage-bau Hambach in Germany.

Modern Surface Mining

When coal seams are near the surface, it may be economical to extract the coal using open cut (also referred to as open cast, open pit, mountaintop removal or strip) mining methods. Open cast coal mining recovers a greater proportion of the coal deposit than underground methods, as more of the coal seams in the strata may be exploited. Large open cast mines can cover an area of many square kilometers and use very large pieces of equipment. This equipment can include the following: Draglines which operate by removing the overburden, power shovels, large trucks in which transport overburden and coal, bucket wheel excavators, and conveyors. In this mining method, explosives are first used in order to break through the surface, or overburden, of the mining area. The overburden is then removed by draglines or by shovel and truck. Once the coal seam is exposed, it is drilled, fractured and thoroughly mined in strips. The coal is then loaded onto large trucks or conveyors for transport to either the coal preparation plant or directly to where it will be used.

Trucks loaded with coal at the Cerrejón coal mine in Colombia

Most open cast mines in the United States extract bituminous coal. In Canada (BC), Australia and South Africa, open cast mining is used for both thermal and metallurgi-cal coals. In New South Wales open casting for steam coal and anthracite is practised. Surface mining accounts for around 80 percent of production in Australia, while in the US it is used for about 67 percent of production. Globally, about 40 percent of coal production involves surface mining.

Strip Mining

Strip mining exposes coal by removing earth above each coal seam. This earth is referred to as overburden and is removed in long strips. The overburden from the first strip is deposited in an area outside the planned mining area and referred to as out-of-pit dumping. Overburden from subsequent strips are deposited in the void left from mining the coal and overburden from the previous strip. This is referred to as in-pit dumping.

It is often necessary to fragment the overburden by use of explosives. This is accomplished by drilling holes into the overburden, filling the holes with explosives, and detonating the explosive. The overburden is then removed, using large earth-moving equipment, such as draglines, shovel and trucks, excavator and trucks, or bucket-wheels and conveyors. This overburden is put into the previously mined (and now empty) strip. When all the overburden is removed, the underlying coal seam will be exposed (a 'block' of coal). This block of coal may be drilled and blasted (if hard) or otherwise loaded onto trucks or conveyors for transport to the coal preparation (or wash) plant. Once this strip is empty of coal, the process is repeated with a new strip being created next to it. This method is most suitable for areas with flat terrain.

Equipment to be used depends on geological conditions. For example, to remove overburden that is loose or unconsolidated, a bucket wheel excavator might be the most productive. The life of some area mines may be more than 50 years.

Contour Mining

The contour mining method consists of removing overburden from the seam in a pattern following the contours along a ridge or around the hillside. This method is most commonly used in areas with rolling to steep terrain. It was once common to deposit the spoil on the downslope side of the bench thus created, but this method of spoil disposal consumed much additional land and created severe landslide and erosion problems. To alleviate these problems, a variety of methods were devised to use freshly cut overburden to refill mined-out areas. These haul-back or lateral movement methods generally consist of an initial cut with the spoil deposited downslope or at some other site and spoil from the second cut refilling the first. A ridge of undisturbed natural material 15 to 20 ft (5–6 m) wide is often intentionally left at the outer edge of the mined area. This barrier adds stability to the reclaimed slope by preventing spoil from slumping or sliding downhill

The limitations of contour strip mining are both economic and technical. When the operation reaches a predetermined stripping ratio (tons of overburden/tons of coal), it is not profitable to continue. Depending on the equipment available, it may not be technically feasible to exceed a certain height of highwall. At this point, it is possible to produce more coal with the augering method in which spiral drills bore tunnels into a highwall laterally from the bench to extract coal without removing the overburden.

Mountaintop Removal Mining

Mountaintop coal mining is a surface mining practice involving removal of mountaintops to expose coal seams, and disposing of associated mining overburden in adjacent "valley fills." Valley fills occur in steep terrain where there are limited disposal alternatives.

Mountaintop removal combines area and contour strip mining methods. In areas with rolling or steep terrain with a coal seam occurring near the top of a ridge or hill, the entire top is removed in a series of parallel cuts. Overburden is deposited in nearby valleys and hollows. This method usually leaves ridge and hill tops as flattened plateaus. The process is highly controversial for the drastic changes in topography, the practice of creating *head-of-hollow-fills*, or filling in valleys with mining debris, and for covering streams and disrupting ecosystems.

Spoil is placed at the head of a narrow, steep-sided valley or hollow. In preparation for filling this area, vegetation and soil are removed and a rock drain constructed down the middle of the area to be filled, where a natural drainage course previously existed. When the fill is completed, this underdrain will form a continuous water runoff system from the upper end of the valley to the lower end of the fill. Typical head-of-hollow fills are graded and terraced to create permanently stable slopes.

Underground Mining

Coal wash plant in Clay County, Kentucky

Most coal seams are too deep underground for opencast mining and require underground mining, a method that currently accounts for about 60 percent of world coal production. In deep mining, the room and pillar or bord and pillar method progresses along the seam, while pillars and timber are left standing to support the mine roof.

Once room and pillar mines have been developed to a stopping point (limited by geology, ventilation, or economics), a supplementary version of room and pillar mining, termed second mining or retreat mining, is commonly started. Miners remove the coal in the pillars, thereby recovering as much coal from the coal seam as possible. A work area involved in pillar extraction is called a pillar section.

Modern pillar sections use remote-controlled equipment, including large hydraulic mobile roof-supports, which can prevent cave-ins until the miners and their equipment have left a work area. The mobile roof supports are similar to a large dining-room table, but with hydraulic jacks for legs. After the large pillars of coal have been mined away, the mobile roof support's legs shorten and it is withdrawn to a safe area. The mine roof typically collapses once the mobile roof supports leave an area.

Remote Joy HM21 Continuous Miner used underground

There are six principal methods of underground mining:

- Longwall mining accounts for about 50 percent of underground production. The longwall shearer has a face of 1,000 feet (300 m) or more. It is a sophisticated machine with a rotating drum that moves mechanically back and forth across a wide coal seam. The loosened coal falls onto an armored chain conveyor or pan line that takes the coal to the conveyor belt for removal from the work area. Longwall systems have their own hydraulic roof supports which advance with the machine as mining progresses. As the longwall mining equipment moves forward, overlying rock that is no longer supported by coal is allowed to fall behind the operation in a controlled manner. The supports make possible high levels of production and safety. Sensors detect how much coal remains in the seam while robotic controls enhance efficiency. Longwall systems allow a 60-to-100 percent coal recovery rate when surrounding geology allows their use. Once the coal is removed, usually 75 percent of the section, the roof is allowed to collapse in a safe manner.

- Continuous mining utilizes a Continuous Miner Machine with a large rotating steel drum equipped with tungsten carbide picks that scrape coal from the seam. Operating in a "room and pillar" (also known as "bord and pillar") system—where the mine is divided into a series of 20-to-30 foot (5–10 m) "rooms" or work areas cut into the coalbed—it can mine as much as 14 tons of coal a minute, more than a non-mechanised mine of the 1920s would produce in an entire day. Continuous miners account for about 45 percent of underground coal production. Conveyors transport the removed coal from the seam. Remote-controlled continuous miners are used to work in a variety of difficult seams and conditions, and robotic versions controlled by computers are becoming increasingly common. Continuous mining is a misnomer, as room and pillar coal mining is very cyclical. In the US, one can generally cut 20 ft or 6 meters (or a bit more with MSHA permission) (12 meters or roughly 40 ft in South Africa before the Continuous Miner goes out and the roof is supported by the Roof Bolter), after which, the face has to be serviced, before it can be advanced again. During servicing, the "continuous" miner moves to another face. Some continuous miners can bolt and rock dust the face (two major components of servicing) while cutting coal, while a trained crew may be able to advance ventilation, to truly earn the "continuous" label. However, very few mines are able to achieve it. Most continuous mining machines in use in the US lack the ability to bolt and dust. This may partly be because incorporation of bolting makes the machines wider, and therefore, less maneuverable.

- Room and pillar mining consists of coal deposits that are mined by cutting a network of rooms into the coal seam. Pillars of coal are left behind in order to keep up the roof. The pillars can make up to forty percent of the total coal in the seam, however where there was space to leave head and floor coal there is evidence from recent open cast excavations that 18th-century operators used a variety of room and pillar techniques to remove 92 percent of the *in situ* coal. However, this can be extracted at a later stage.

- Blast mining or conventional mining, is an older practice that uses explosives such as dynamite to break up the coal seam, after which the coal is gathered and loaded onto shuttle cars or conveyors for removal to a central loading area. This process consists of a series of operations that begins with "cutting" the coalbed so it will break easily when blasted with explosives. This type of mining accounts for less than 5 percent of total underground production in the US today.

- Shortwall mining, a method currently accounting for less than 1 percent of deep coal production, involves the use of a continuous mining machine with movable roof supports, similar to longwall. The continuous miner shears coal panels 150 to 200 feet (40 to 60 m) wide and more than a half-mile (1 km) long, having regard to factors such as geological strata.

- Retreat mining is a method in which the pillars or coal ribs used to hold up the mine roof are extracted; allowing the mine roof to collapse as the mining works back towards the entrance. This is one of the most dangerous forms of mining, owing to imperfect predictability of when the roof will collapse and possibly crush or trap workers in the mine.

Production

Coal is mined commercially in over 50 countries. Over 7,036 Mt/yr of hard coal is currently produced, a substantial increase over the past 25 years. In 2006, the world production of brown coal and lignite was slightly over 1,000 Mt, with Germany the world's largest brown coal producer at 194.4 Mt, and China second at 100.6 Mt.

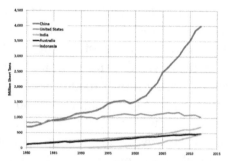

Coal production trends 1980-2012 in the top five coal-producing countries (US EIA)

Coal mine in Australia

Coal mine in China

Coal production has grown fastest in Asia, while Europe has declined. The top coal mining nations (figures in brackets are 2009 estimate of total coal production in millions of tons) are:

- China (3,050 Mt)

- United States (973 Mt)

- India (557 Mt)

- Australia (409 Mt)

- Russia (298 Mt)

- Indonesia (252 Mt)

- South Africa (250 Mt)

- Poland (135 Mt)

- Kazakhstan (101 Mt)

- Colombia (75 Mt)

Most coal production is used in the country of origin, with around 16 percent of hard coal production being exported.

Global coal production is expected to reach 7,000 Mt/yr in 2030 (Update required, world coal production is already past 7,000 Mt/yr and by 2030 will probably be closer to 13,000 Mt/yr), with China accounting for most of this increase. Steam coal production is projected to reach around 5,200 Mt/yr; coking coal 620 Mt/yr; and brown coal 1,200 Mt/yr.

Coal reserves are available in almost every country worldwide, with recoverable reserves in around 70 countries. At current production levels, proven coal reserves are estimated to last 147 years. However, production levels are by no means level, and are in fact increasing and some estimates are that peak coal could arrive in many countries such as China and America by around 2030. Coal reserves are usually stated as either (1) "Resources" ("measured" + "indicated" + "inferred" = "resources", and then, a smaller number, often only 10-20% of "resources," (2) "Run of Mine" (ROM) reserves, and finally (3) "marketable reserves", which may be only 60% of ROM reserves. The standards for reserves are set by stock exchanges, in consultation with industry associations. For example, in ASEAN countries reserves standards follow the Australasian Joint Ore Reserves Committee Code (JORC) used by the Australian Securities Exchange.

Modern Mining

Laser profiling of a minesite by a coal miner using a Maptek I-site laser scanner in 2014

Technological advancements have made coal mining today more productive than it has ever been. To keep up with technology and to extract coal as efficiently as possible modern mining personnel must be highly skilled and well trained in the use of complex, state-of-the-art instruments and equipment. Many jobs require four-year university degrees. Computer knowledge has also become greatly valued within the industry as most of the machines and safety monitors are computerized.

The use of sophisticated sensing equipment to monitor air quality is common and has replaced the use of small animals such as canaries, often referred to as "miner's canaries".

In the United States, the increase in technology has significantly decreased the mining workforce.The average number of employees at U.S. coal mines decreased 12.0% to 65,971 employees, the lowest on record since EIA began collecting data in 1978. As the number of coal workers has decreased, a study caused controversy as GreenTech Media pointed out that investment for the retraining all coal workers for solar jobs was modest - one year of coal CEO pay could retrain every US miner to work in the solar industry.

Safety

Dangers to Miners

The Farmington coal mine disaster kills 78. West Virginia, US, 1968.

Historically, coal mining has been a very dangerous activity and the list of historical coal mining disasters is a long one. In the US alone, more than 100,000 coal miners were killed in accidents in the twentieth century, 90 percent of the fatalities occurring in the first half of the century. More than 3,200 died in 1907 alone.

Open cut hazards are principally mine wall failures and vehicle collisions; underground mining hazards include suffocation, gas poisoning, roof collapse, rock burst, outbursts, and gas explosions.

Firedamp explosions can trigger the much-more-dangerous coal dust explosions, which can engulf an entire pit. Most of these risks can be greatly reduced in modern mines, and multiple fatality incidents are now rare in some parts of the developed world. Modern mining in the US results in approximately 30 deaths per year due to mine accidents.

However, in lesser developed countries and some developing countries, many miners continue to die annually, either through direct accidents in coal mines or through adverse health consequences from working under poor conditions. China, in particular, has the highest number of coal mining related deaths in the world, with official statistics claiming that 6,027 deaths occurred in 2004. To compare, 28 deaths were reported in the US in the same year. Coal production in China is twice that in the US, while the number of coal miners is around 50 times that of the US, making deaths in coal mines in China 4 times as common per worker (108 times as common per unit output) as in the US.

Mine disasters have still occurred in recent years in the US, Examples include the Sago Mine disaster of 2006, and the 2007 mine accident in Utah's Crandall Canyon Mine, where nine miners were killed and six entombed. In the decade 2005-2014, US coal mining fatalities averaged 28 per year. The most fatalities during the 2005-2014 decade were 48 in 2010, the year of the Upper Big Branch Mine disaster in West Virginia, which killed 29 miners.

Miners can be regularly monitored for reduced lung function due to coal dust exposure using spirometry.

Chronic lung diseases, such as pneumoconiosis (black lung) were once common in miners, leading to reduced life expectancy. In some mining countries black lung is still common, with 4,000 new cases of black lung every year in the US (4 percent of workers annually) and 10,000 new cases every year in China (0.2 percent of workers). Rates may be higher than reported in some regions.

Build-ups of a hazardous gas are known as damps, possibly from the German word "Dampf" which means steam or vapor:

- Black damp: a mixture of carbon dioxide and nitrogen in a mine can cause suffocation, and is formed as a result of corrosion in enclosed spaces so removing oxygen from the atmosphere.

- After damp: similar to black damp, after damp consists of carbon monoxide, carbon dioxide and nitrogen and forms after a mine explosion.

- Fire damp: consists of mostly methane, a highly flammable gas that explodes between 5% and 15% - at 25% it causes asphyxiation.

- Stink damp: so named for the rotten egg smell of the hydrogen sulphide gas, stink damp can explode and is also very toxic.

- White damp: air containing carbon monoxide which is toxic, even at low concentrations

Safer Times in Modern Mining

Improvements in mining methods (e.g. longwall mining), hazardous gas monitoring (such as safety-lamps or more modern electronic gas monitors), gas drainage, electrical equipment, and ventilation have reduced many of the risks of rock falls, explosions, and unhealthy air quality. Gases released during the mining process can be recovered to generate electricity and improve worker safety with gas engines. Another innovation in recent years is the use of closed circuit escape respirators, respirators that contain oxygen for situations where mine ventilation is compromised. Statistical analyses performed by the US Department of Labor's Mine Safety and Health Administration (MSHA) show that between 1990 and 2004, the industry cut the rate of injuries by more than half and fatalities by two-thirds. However, according to the Bureau of Labor Statistics, even in 2006, mining remained the second most dangerous occupation in America, when measured by fatality rate. However, these numbers include all mining, with oil and gas mining contributing the majority of fatalities; coal mining resulted in only 47 fatalities that year.

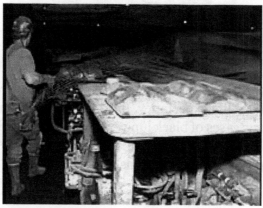

A video on the use of roof screens in underground coal mines

Environmental Impacts

Coal mining can result in a number of adverse effects on the environment.

Surface mining of coal completely eliminates existing vegetation, destroys the genetic soil profile, displaces or destroys wildlife and habitat, degrades air quality, alters current land uses, and to some extent permanently changes the general topography of the area mined. This often results in a scarred landscape with no scenic value. Of greater concern, the movement, storage, and redistribution of soil during mining can disrupt the community of soil microorganisms and consequently nutrient cycling processes. Rehabilitation or reclamation mitigates some of these concerns and is required by US Federal Law, specifically the Surface Mining Control and Reclamation Act of 1977.

Mine dumps (tailings) could produce acid mine drainage which can seep into waterways and aquifers, with consequences on ecological and human health.

If underground mine tunnels collapse, they cause subsidence of the ground above. Subsidence can damage buildings, and disrupt the flow of streams and rivers by interfering with the natural drainage.

During actual mining operations, methane, a known greenhouse gas, may be released into the air.

Coal Mining by Country

A view of Murton colliery near Seaham, United Kingdom, 1843

Top 10 hard and brown coal producers in 2012 were (in million metric tons): China 3,621, United States 922, India 629, Australia 432, Indonesia 410, Russia 351, South Africa 261, Germany 196, Poland 144, and Kazakhstan 122.

Australia

Coal is mined in every state of Australia, but mainly in Queensland, New South Wales and Victoria. It is mostly used to generate electricity, and 75% of annual coal production is exported, mostly to eastern Asia.

In 2007, 428 million tonnes of coal was mined in Australia. In 2007, coal provided about 85% of Australia's electricity production. In fiscal year 2008/09, 487 million tonnes of coal was mined, and 261 million tonnes was exported. In fiscal year 2013/14, 430.9 million tonnes of coal was mined, and 375.1 million tonnes was exported. In 2013/14, coal provided about 69% of Australia's electricity production.

In 2013, Australia was the world's fifth-largest coal producer, after China, the United States, India, and Indonesia. However, in terms of proportion of production exported, Australia is the world's second largest coal exporter, as it exports roughly 73% of its coal production. Indonesia exports about 87% of its coal production.

Canada

Canada was ranked as the 15th coal producing country in the world in 2010, with a total production of 67.9 million tonnes. Canada's coal reserves, the 12th largest in the world, are located largely in the province of Alberta.

The first coal mines in North America were located in Joggins and Port Morien, Nova Scotia, mined by French settlers beginning in the late 1600s. The coal was used for the British garrison at Annapolis Royal, and in construction of the Fortress of Louisbourg.

Chile

China

The People's Republic of China is by far the largest producer of coal in the world, producing over 2.8 billion tons of coal in 2007, or approximately 39.8 percent of all coal produced in the world during that year. For comparison, the second largest producer, the United States, produced more than 1.1 billion tons in 2007. An estimated 5 million people work in China's coal-mining industry. As many as 20,000 miners die in accidents each year. Most Chinese mines are deep underground and do not produce the surface disruption typical of strip mines. Although there is some evidence of reclamation of mined land for use as parks, China does not require extensive reclamation and is creating significant acreages of abandoned mined land, which is unsuitable for agriculture or other human uses, and inhospitable to indigenous wildlife. Chinese underground mines often experience severe surface subsidence (6–12 meters), negatively impacting farmland because it no longer drains well. China uses some subsidence areas for aquaculture ponds but has more than they need for that purpose. Reclamation of subsided ground is a significant problem in China. Because most Chinese coal is for domestic consumption, and is burned with little or no air pollution control equipment, it contributes greatly to visible smoke and severe air pollution in industrial areas using coal for fuel. China's total energy uses 67% from coal mines.

Colombia

Opencast coal mine at Cerrejón

Some of the world's largest coal reserves are located in South America, and an opencast mine at Cerrejón in Colombia is one of the world's largest open pit mines. Output of the mine in 2004 was 24.9 million tons (compared to total global hard coal production of 4,600 million tons). Cerrejón contributed about half of Colombia's coal exports of 52 million tons that year, with Colombia ranked sixth among major coal exporting nations. The company planned to expand production to 32 million tons by 2008. The company has its own 150 km standard-gauge railroad, connecting the mine to its coal-loading terminal at Puerto Bolívar on the Caribbean coast. There are two 120-car unit trains, each carrying 12,000 tons of coal per trip. The round-trip time for each train, including loading and unloading, is about 12 hours. The coal facilities at the port are capable of loading 4,800 tons per hour onto vessels of up to 175,000 tons of dead weight. The mine, railroad and port operate 24 hours per day. Cerrejón directly employs 4,600 workers, with a further 3,800 employed by contractors. The reserves at Cerrejón are low-sulfur, low-ash, bituminous coal. The coal is mostly used for electric power generation, with some also used in steel manufacture. The surface mineable reserves for the current contract are 330 million tons. However, total proven reserves to a depth of 300 metres are 3,000 million tons.

Germany

Germany has a long history of coal mining, going back to the Middle Ages. Coal mining greatly increased during the industrial revolution and the following decades. The main mining areas were around Aachen, the Ruhr and Saar area, along with many smaller areas in other parts of Germany. These areas grew and were shaped by coal mining and coal processing, and this is still visible even after the end of the coal mining.

3,500-4,000 environmental activists blocking a coal mine to limit climate change (Ende Gelände 2016).

Coal mining reached its peak in the first half of the 20th century. After 1950, the coal producers started to struggle financially. In 1975, a subsidy was introduced (*Kohlepfennig*). In 2007, the Bundestag decided to end subsidies by 2018. As a consequence, RAG Aktiengesellschaft, the owner of the two remaining coal mines in Germany, announced it would close all mines by 2018, thus ending coal mining in Germany.

India

Coal mining in India has a long history of commercial exploitation covering nearly 220 years starting in 1774 with John Sumner and Suetonius Grant Heatly of the East India Company in the Raniganj Coalfield along the Western bank of river Damodar. However, for about a century the growth of Indian coal mining remained sluggish for want of demand but the introduction of steam locomotives in 1853 gave a fillip to it. Within a short span, production rose to an annual average of 1 million tonne (mt) and India could produce 6.12 mts. per year by 1900 and 18 mts per year by 1920. The production got a sudden boost from the First World War but went through a slump in the early thirties. The production reached a level of 29 mts. by 1942 and 30 mts. by 1946. With the advent of Independence, the country embarked upon the 5-year development plans. At the beginning of the 1st Plan, annual production went up to 33 mts. During the 1st Plan period itself, the need for increasing coal production efficiently by systematic and scientific development of the coal industry was being felt. Setting up of the National Coal Development Corporation (NCDC), a Government of India Undertaking in 1956 with the collieries owned by the railways as its nucleus was the first major step towards planned development of Indian Coal Industry. Along with the Singareni Collieries Company Ltd. (SCCL) which was already in operation since 1945 and which became a Government company under the control of Government of Andhra Pradesh in 1956, India thus had two Government coal companies in the fifties. SCCL is now a joint undertaking of Government of Telangana and Government of India sharing its equity in 51:49 ratio.

Japan

The Japanese archipelago counts four main islands, the richest coal deposits have been found on the northernmost and the southernmost island: Hokkaidō and Kyūshū.

The Daikōdō (大抗道), the first adit of the Horonai mine (1879).(also known as the Otowakō (音羽坑))

Japan has a long history of coal mining dating back into the Japanese Middle Ages. It has been told that the first coal has been discovered by a farmer couple in the region of Ōmuta, central Kyūshū in 1469. Nine years later, in 1478, local farmers discovered burning stones in the north of the Island, which meant the start of the exploitation of the Chikuhō coalfield.

The discovery of the coalfields in the north are only since the Japanese industrialization. One of the first mines in Hokkaidō was the Hokutan Horonai coal mine.

Poland

Russia

Russia was ranked as the 5th coal producing country in the world in 2010, with a total production of 316.9 million tonnes. Russia itself is the possessor of the world's second largest coal reserves. Russia also has equal rights to coal located in the Arctic archipelago of Svalbard, in accordance with the Svalbard Treaty.

Spain

Spain was ranked as the 30th coal producing country in the world in 2010. The coal miners of Spain were active in the Spanish Civil War on the Republican side. In October 1934, in Asturias, union miners and others suffered a fifteen-day siege in Oviedo and Gjion. There is a museum dedicated to coal mining in the region of Catalonia, called Cercs Mine Museum.

South Africa

South Africa is one of the ten largest coal producing and the fourth largest coal exporting country in the world.

Taiwan

In Taiwan, coal is distributed mainly in the northern area. All of the commercial coal deposits occurred in three Miocene coal-bearing formations, which are the Upper, the Middle and the Lower Coal Measures. The Middle Coal Measures was the most important with its wide distribution, great number of coal beds and extensive potential reserves. Taiwan has coal reserves estimated to be 100-180 Mt. However, coal output had been small, amounting to 6,948 metric tonnes per month from 4 pits before it ceased production effectively in 2000. The abandoned coal mine in Pingxi District, New Taipei has now turned into the Taiwan Coal Mine Museum.

Abandoned coal mine in Pingxi, New Taipei.

Ukraine

In 2012 coal production in Ukraine amounted to 85.946 million tonnes, up 4.8% from 2011. Coal consumption that same year grew by to 61.207 million tonnes, up 6.2% compared with 2011.

More than 90 percent of Ukraine's coal production comes from the Donets Basin. The country's coal industry employs about 500,000 people. Ukrainian coal mines are among the most dangerous in the world, and accidents are common. Furthermore, the country is plagued with extremely dangerous illegal mines.

United Kingdom

The United Kingdom was ranked as the 24th coal producing country in the world in 2010, with a total production of 18.2 million tonnes. Coal mining in the United Kingdom probably dates to Roman times; coal production increased significantly during the Industrial Revolution in the 19th century and peaked during World War I. As a result of its long history with coal Britain's economically recoverable coal reserves have decreased, and more than twice as much coal is now imported than produced.

United States

Miners at the Virginia-Pocahontas Coal Company Mine in 1974

The American share of world coal production remained steady at about 20 percent from 1980 to 2005, at about 1 billion short tons per year. The United States was ranked as the 2nd coal producing country in the world in 2010, and possesses the largest coal reserves in the world. In 2008 then-President George W. Bush stated that coal was the most reliable source of electricity. However, in 2011 President Barack Obama said that the US should rely more on "clean" sources of energy that emit lower or no carbon dioxide pollution. As of 2013, while domestic coal consumption for electric power was being displaced by natural gas, exports were increasing. US coal production increasingly comes from strip mines in the western United States, such as from the Powder River Basin in Wyoming and Montana.

Coal has come under continued price pressure from natural gas and renewable energy sources, which has resulted in a rapid decline of coal in the U.S. and several notable bankruptcies including Peabody Energy. On April 13, 2016 it reported, its revenue tumbled 17 percent as coal price fell and lost 2 billion dollars on the previous year. It then filed Chapter 11 bankruptcy on April 13, 2016. The Harvard Business Review discussed retraining coal workers for solar photovoltaic employment because of the rapid rise in U.S. solar jobs. A recent study indicated that this was technically possible and would account for only 5% of the industrial revenue from a single year to provide coal workers with job security in the energy industry as whole.

References

- Hulse, David H: The Early Development of the Steam Engine; TEE Publishing, Leamington Spa, UK, 1999 ISBN 1-85761-107-1.

- Hills, Rev. Dr. Richard (2006), James Watt Vol 3: Triumph through Adversity, 1785-819, Ashbourne, Derbyshire, England: Landmark Publishing, p. 217, ISBN 1-84306-045-0.

- Benett, Stuart (1986). A History of Control Engineering 1800-1930. Institution of Engineering and Technology. ISBN 0863410472.

- Thompson, Ross (2009). Structures of Change in the Mechanical Age: Technological Invention in the United States 1790-1865. Baltimore, MD: The Johns Hopkins University Press. ISBN 978-0-8018-9141-0.

- Thomson, Ross (1989). The Path to Mechanized Shoe Production in the United States. University of North Carolina Press. ISBN 978-0807818671.

- Bidwell, John (2013). American Paper Mills, 1690-1832: A Directory of the Paper Trade with Notes... Dartmouth College Press. pp. 154–155. ISBN 978-1-58465-964-8.

- Misa, Thomas J. (1995). A Nation of Steel: The Making of Modern America 1965–1925. Baltimore and London: Johns Hopkins University Press. p. 243. ISBN 978-0-8018-6502-2.

- Paper Machine Clothing: Key to the Paper Making Process Sabit Adanur, Asten,CRC Press, 1997, p. 120–136, ISBN 978-1-56676-544-2.

- "Jobs to go as South West Wales coal mine is mothballed". South Wales Evening Post. 2015-06-26. Retrieved 2016-08-20.

- Riley, Charles; Isidore, Chris. "Top U.S. coal company Peabody Energy files for bankruptcy". CNNMoney. Retrieved 2016-04-13.

Permissions

Index

www.ingramcontent.com/pod-product-compliance
Lightning Source LLC
Jackson TN
JSHW052151130125
77033JS00004B/172